Student Practice Sheets

Prealgebra

FIFTH EDITION

Richard N. Aufmann
Palomar College

Vernon C. Barker
Palomar College

Joanne S. Lockwood
New Hampshire Community Technical College

Prepared by

Christine S. Verity

Y0-BPX-401

BROOKS/COLE
CENGAGE Learning

Australia • Brazil • Japan • Korea • Mexico • Singapore • Spain • United Kingdom • United States

For product information and technology assistance, contact us at
**Cengage Learning Customer & Sales Support,
1-800-354-9706**

For permission to use material from this text or product, submit all requests online at **www.cengage.com/permissions**
Further permissions questions can be emailed to
permissionrequest@cengage.com

ISBN-13: 978-0-547-20891-6
ISBN-10: 0-547-20891-X

Brooks/Cole
20 Davis Drive
Belmont, CA 94002-3098
USA

Cengage Learning is a leading provider of customized learning solutions with office locations around the globe, including Singapore, the United Kingdom, Australia, Mexico, Brazil, and Japan. Locate your local office at: **www.cengage.com/global**

Cengage Learning products are represented in Canada by Nelson Education, Ltd.

To learn more about Brooks/Cole, visit
www.cengage.com/brookscole

Purchase any of our products at your local college store or at our preferred online store
www.ichapters.com

Printed in the United States of America
1 2 3 4 5   14 13 12 11 10

ED011

Contents

Name   Score

Graph the number on the number line.

1. 2

2. 6

1. _____

2. _____

3. 7

4. 1

3. _____

4. _____

Place the correct symbol, < or >, between the two numbers.

5. 54   47   **6.** 29   11   **7.** 96   97

5. _____

6. _____

7. _____

8. 9   13   **9.** 391   250   **10.** 1010   1110

8. _____

9. _____

10. _____

11. 1   64   **12.** 105   0   **13.** 666   777

11. _____

12. _____

13. _____

14. 32,000   3200   **15.** 5910   5901   **16.** 63,528   64,249

14. _____

15. _____

16. _____

Write the given numbers in order from smallest to largest.

17. 57, 81, 6, 21, 8, 80   **18.** 64, 14, 19, 72, 6   **19.** 312, 435, 812, 200, 814

17. _____

18. _____

19. _____

1

Name   Score

Write the number in words.

1. 862   2. 308   3. 5125   1. _____

                                                         2. _____

                                                         3. _____

4. 380,751   5. 800,001   6. 7,640,723   4. _____

                                                            _____

                                                         5. _____

                                                            _____

                                                         6. _____

                                                            _____

Write the number in standard form.

7. Thirty-three   8. Two hundred seventy-four   7. _____

                                                              8. _____

9. Nine thousand five hundred   10. Fifty-six thousand three hundred   9. _____
    twenty-seven   twenty
                                                             10. _____

11. Four hundred sixty thousand 12. Four million twelve thousand   11. _____
    three hundred three   nine hundred eighty-six
                                                             12. _____

13. One million five   14. Eight million one thousand fifty   13. _____

                                                             14. _____

Write the number in expanded form.

15. 256   16. 4703   15. _____

                                       16. _____

17. 50,916   18. 500,400   17. _____

                                       18. _____

19. 920,010   20. 4,271,020   19. _____

                                       20. _____

Name   Score

Round the number to the given place value.

1.	747	Tens	2.	711	Tens

1. _____

2. _____

3.	361	Hundreds	4.	589	Hundreds

3. _____

4. _____

5.	1050	Hundreds	6.	7149	Hundreds

5. _____

6. _____

7.	4461	Thousands	8.	9620	Thousands

7. _____

8. _____

9.	74,510	Thousands	10.	69,372	Thousands

9. _____

10. _____

11.	250,650	Ten-thousands	12.	841,512	Ten-thousands

11. _____

12. _____

13.	3,467,000	Millions	14.	7,600,475	Millions

13. _____

14. _____

Name _____   Score _____

Solve.

1. Your mathematics textbook has 863 pages. Your friend's mathematics textbook has 770 pages. Which textbook has more pages, yours or your friend's?

2. A BK Veggie burger contains 420 calories. A Burger King Whopper contains 670 calories. Which sandwich contains less calories, the Veggie Burger or the Whopper?

3. One light bulb has 60 watts. Another light bulb has 100 watts. Which light bulb has more watts?

4. A premium mattress has 780 coils. A regular mattress has 390 coils. Which mattress has a smaller number of coils?

5. The area of Montana, including water, is 147,042 square miles. The area of New Mexico, including water, is 121,593 square miles. Which state has the larger area?

6. During a bobsled run at the Olympics, a team reached a speed of 64 mph in the middle of the run and 83 mph at the end of the run. At which point was the team going slower, the middle or the end?

7. The Appalachian Trail is 2160 miles long. What is the length of the Appalachian Trail to the nearest hundred miles?

8. The area of New Hampshire, including water, is 9351 square miles. What is the area of New Hampshire rounded to the nearest thousand miles?

9. For 2005, Chevron reported record profits of $14,100,000,000. Round this number to the nearest billion dollar.

10. The Italian national anthem was written in 1847. The Denmark national anthem was written in 1844. Which national anthem was written first, Denmark's or Italy's?

1. _____

2. _____

3. _____

4. _____

5. _____

6. _____

7. _____

8. _____

9. _____

10. _____

4

Name   Score

Add.

1.	391 + 475	**2.**	568 + 284	**3.**	821 + 1774

1. _____

2. _____

3. _____

4.	8515 5307 + 4518	**5.**	7810 9537 + 3284	**6.**	9478 3925 + 6227

4. _____

5. _____

6. _____

7. $567 + 23 + 849$ **8.** $723 + 96 + 452$

7. _____

8. _____

9. $35 + 8701 + 65,247 + 765$ **10.** $48 + 7449 + 26,835 + 98$

9. _____

10. _____

Evaluate the variable expression $x + y$ for the given values of x and y.

11. $x = 354; y = 742$ **12.** $x = 75,206; y = 2406$ **13.** $x = 8143; y = 1987$

11. _____

12. _____

13. _____

Evaluate the variable expression $a + b + c$ for the given values of a, b and c.

14. $a = 387; b = 907; c = 24$ **15.** $a = 2761; b = 5789;$
$c = 948$ **16.** $a = 8145; b = 67,572;$
$c = 300$

14. _____

15. _____

16. _____

Solve.

17. Is 14 a solution of the equation $x + 5 = 9$?   **18.** Is 27 a solution to the equation $38 = 9 + y$?   **19.** Is 34 a solution to the equation $7 + h = 41$?

17. _____

18. _____

19. _____

Name   Score

Subtract.

1. 179 − 84
2. 453 − 53
3. 164 − 93

4. 8714 − 301
5. 1539 − 722
6. 2408 − 204

7. 3645 − 376
8. 5094 − 695
9. 1419 − 453

10. 22,653 − 18,734
11. 53,417 − 26,718
12. 79,600 − 35,752

13. 83,602 − 64,575
14. 248,000 − 96,729
15. 900,721 − 546,847

1. _____
2. _____
3. _____
4. _____
5. _____
6. _____
7. _____
8. _____
9. _____
10. _____
11. _____
12. _____
13. _____
14. _____
15. _____

Evaluate the variable expression $x - y$ for the given values of x and y.

16. $x = 7538; y = 532$
17. $x = 1249; y = 828$
18. $x = 9527; y = 7416$
19. $x = 7884; y = 1733$
20. $x = 30,736; y = 5947$
21. $x = 64,057; y = 7316$

16. _____
17. _____
18. _____
19. _____
20. _____
21. _____

Solve.

22. Is 56 a solution of the equation $x - 15 = 41$?
23. Is 42 a solution to the equation $6 = 56 - y$?
24. Is 18 a solution to the equation $39 - h = 21$?

22. _____
23. _____
24. _____

Name _____   Score _____

Solve.

1. You paid $67 for a coat and $14 for a hat. Find the total cost of the coat and hat.

2. A homemaker has a monthly budget of $225 for food, $65 for car expenses, and $50 for entertainment. Find the total amount budgeted for the three items each month.

1. _____

2. _____

3. Your monthly food budget is $380. How much is left in the food budget after spending $169 on groceries?

4. You had a bank balance of $932. You then wrote checks for $183, $74, and $25. Find the new bank balance.

3. _____

4. _____

5. Your monthly budget for household expenses is $800. After $238 is spent for food, $129 for clothes, and $48 for entertainment, how much is left in the budget?

6. How much larger is Texas than Wyoming? Texas is 266,873 square miles in area, and Wyoming is 97,809 square miles in area.

5. _____

6. _____

7. The attendance at the Thursday afternoon lecture was 2592, and the attendance at the Friday afternoon lecture was 643. Find the total attendance for the two lectures.

8. A hospital's emergency staff treated 83 people on Friday, 92 people on Saturday, and 64 people on Sunday. How many people did the emergency room staff treat from Friday to Sunday?

7. _____

8. _____

9. You had a balance of $753 in your checking account before making deposits of $158, $269, and $374. What is your new checking account balance?

10. You are leaving for a four-day vacation. The odometer on your car reads 61,795 miles You plan to drive 185 miles the first day, 209 miles the second day, and 174 miles the third day. What will be your odometer reading at the end of the trip?

9. _____

10. _____

11. How much change will a customer receive after paying for a $27 purchase with a $50 bill?

12. The down payment on a boat costing $9655 is $1931. Find the amount that remains to be paid.

11. _____

12. _____

7

Name   Score

Multiply.

1.	$\begin{array}{r} 14 \\ \times\ 28 \end{array}$	2.	$\begin{array}{r} 30 \\ \times\ 62 \end{array}$	3.	$\begin{array}{r} 75 \\ \times\ 42 \end{array}$

1. _____

2. _____

3. _____

4.	$\begin{array}{r} 281 \\ \times\ 64 \end{array}$	5.	$\begin{array}{r} 495 \\ \times\ 37 \end{array}$	6.	$\begin{array}{r} 845 \\ \times\ 56 \end{array}$

4. _____

5. _____

6. _____

7.	$\begin{array}{r} 391 \\ \times\ 465 \end{array}$	8.	$\begin{array}{r} 724 \\ \times\ 831 \end{array}$	9.	$\begin{array}{r} 298 \\ \times\ 665 \end{array}$

7. _____

8. _____

9. _____

10.	$\begin{array}{r} 3600 \\ \times\ 240 \end{array}$	11.	$\begin{array}{r} 6412 \\ \times\ 180 \end{array}$	12.	$\begin{array}{r} 8923 \\ \times\ 336 \end{array}$

10. _____

11. _____

12. _____

Evaluate the expression for the given values of the variables.

13. $8x$, when $x = 38$

14. $3xy$, when $x = 4$ and $y = 6$

13. _____

14. _____

15. $4ab$, when $a = 59$ and $b = 27$

16. abc, when $a = 5$, $b = 3$, and $c = 10$

15. _____

16. _____

Solve.

17. Is 9 a solution of the equation $6x = 54$?

18. Is 8 a solution of the equation $2x = 32$?

17. _____

18. _____

19. Is 21 a solution of the equation $64 = 3x$?

20. Is 13 a solution of the equation $4x = 52$?

19. _____

20. _____

Name   Score

Write the number in exponential form.

1. $4 \cdot 4 \cdot 4 \cdot 4$

2. $8 \cdot 8 \cdot 8 \cdot 8 \cdot 8 \cdot 8$

3. $3 \cdot 3 \cdot 3 \cdot 3 \cdot 4 \cdot 4$

4. $5 \cdot 5 \cdot 5 \cdot 5 \cdot 7 \cdot 7 \cdot 7$

5. $2 \cdot 2 \cdot 2 \cdot 3 \cdot 3 \cdot 3 \cdot 3 \cdot 7 \cdot 7$

6. $5 \cdot 6 \cdot 7 \cdot 7 \cdot 8 \cdot 8 \cdot 8$

7. $4 \cdot 4 \cdot 5 \cdot 5 \cdot 5 \cdot 6 \cdot 6 \cdot 6 \cdot 7$

8. $7 \cdot 7 \cdot 7 \cdot 15 \cdot 15 \cdot 19 \cdot 19$

1. _____

2. _____

3. _____

4. _____

5. _____

6. _____

7. _____

8. _____

Evaluate.

9. 2^4

10. 3^3

11. 4^3

12. $0^5 \cdot 3^2 \cdot 6^4$

13. $2 \cdot 4^2 \cdot 10$

14. $5^2 \cdot 10^2 \cdot 15$

15. $3 \cdot 10^2 \cdot 7^3$

16. $2^3 \cdot 3^3 \cdot 4^3$

17. $7^2 \cdot 10 \cdot 3^4$

9. _____

10. _____

11. _____

12. _____

13. _____

14. _____

15. _____

16. _____

17. _____

Evaluate the expression for the given values of the variables.

18. $x^4 \cdot y^5$,
    when $x = 2$ and $y = 3$

19. $a^2 \cdot b^3$,
    when $a = 4$ and $b = 5$

20. $m^2 \cdot n^3$,
    when $m = 7$ and $n = 8$

21. $a^2 \cdot b^3 \cdot c^3$,
    when $a = 9$, $b = 2$,
    and $c = 6$

22. $x^3 \cdot y^4 \cdot z^3$,
    when $x = 5$, $y = 0$,
    and $z = 8$

23. $d^2 \cdot e^6 \cdot f$,
    when $d = 5$, $e = 2$,
    and $f = 9$

18. _____

19. _____

20. _____

21. _____

22. _____

23. _____

Name _____   Score _____

Divide.

1. $73 \div 4$	2. $48 \div 7$	3. $40 \div 9$	1. _____
			2. _____
			3. _____
4. $3211 \div 13$	5. $4968 \div 8$	6. $1938 \div 34$	4. _____
			5. _____
			6. _____
7. $693 \div 4$	8. $355 \div 7$	9. $890 \div 6$	7. _____
			8. _____
			9. _____
10. $24,652 \div 3$	11. $59,403 \div 4$	12. $72,641 \div 8$	10. _____
			11. _____
			12. _____

Evaluate the expression $\frac{x}{y}$ for the given values of x and y.

13. $x = 49; y = 7$ 14. $x = 50; y = 0$ 13. _____

15. $x = 7611; y = 3$ 16. $x = 13; y = 1$ 14. _____

15. _____

16. _____

Solve.

17. Is 6 a solution of the equation $\frac{54}{y} = 9$?   18. Is 63 a solution of the equation $\frac{x}{3} = 21$?   17. _____

18. _____

19. Is 13 a solution of the equation $\frac{106}{8} = z$?   20. Is 18 a solution of the equation $\frac{196}{a} = 11$?   19. _____

20. _____

Name   Score

Find all the factors of the number.

1. 15   **2.** 23   **3.** 40   **1.** _____

                                                              2. _____

                                                              3. _____

4. 62   **5.** 81   **6.** 51   **4.** _____

                                                              5. _____

                                                              6. _____

7. 88   **8.** 63   **9.** 35   **7.** _____

                                                              8. _____

                                                              9. _____

10. 45   **11.** 78   **12.** 55   **10.** _____

                                                              11. _____

                                                              12. _____

Find the prime factorization.

13. 8   **14.** 30   **15.** 53   **13.** _____

                                                              14. _____

                                                              15. _____

16. 11   **17.** 20   **18.** 82   **16.** _____

                                                              17. _____

                                                              18. _____

19. 35   **20.** 44   **21.** 69   **19.** _____

                                                              20. _____

                                                              21. _____

22. 72   **23.** 88   **24.** 94   **22.** _____

                                                              23. _____

                                                              24. _____

Name _____   Score _____

Solve.

1. The Environmental Protection Agency (EPA) estimates that a motorcycles gets 38 miles on one gallon of gas. How many miles could it get on 6 gallons of gas?

2. The truck driver drove at a constant speed of 53 miles per hour for 6 hours. Find the distance the truck driver traveled.

1. _____

2. _____

3. A lottery prize of $737,000 is divided equally among 4 winners. What amount does each winner receive?

4. The new movie premiered at 14 theaters last Friday. If 2240 people attended the premiere and they were distributed equally among the 14 theaters, how many people attended each theater?

3. _____

4. _____

5. A librarian catalogued 27 shelves of books. Each shelf held 32 books. How many books did the librarian catalogue?

6. An investor receives a check for $394 each month. How much will the investor receive over a 24-month period?

5. _____

6. _____

7. The total cost of a car, including finance charges, is $7632. This amount is to be repaid in 36 equal monthly payments. What is the amount of each payment?

8. A management consultant received a check for $1755 for 45 hours of work. What is the consultant's hourly wage?

7. _____

8. _____

9. It takes approximately 709 hours for the moon to make one revolution around Earth. How many hours would it take the moon to make 12 revolutions around Earth?

10. A baker can buy 1000 pounds of flour for $150 and one 100-pound bag of sugar for $32. The baker orders 1000 pounds of flour and fifteen 100-pound bags of sugar. What is the total cost of the order?

9. _____

10. _____

11. A student obtains a no-interest loan of $4600 per year for three years. After that time the student must pay off the loan in equal monthly payments for a period of 120 months. What is the amount of the student's monthly payments?

12. A consumer makes a down payment of $4516 on a video entertainment center costing $6724. The balance is to be paid in 12 equal monthly payments. What is the payment for one month?

11. _____

12. _____

Name   Score

Solve.

1. $x + 5 = 11$

2. $5 + n = 8$

3. $x + 7 = 7$

4. $t - 5 = -3$

5. $v - 6 = -2$

6. $x - 4 = -1$

7. $1 - x = 0$

8. $y - 3 = 7$

9. $z - 1 = 1$

10. $x + 13 = 57$

11. $9 + c = 41$

12. $24 = q + 11$

13. $2x = 8$

14. $5x = 25$

15. $54 = 9y$

16. $-20 = -5y$

17. $21r = 21$

18. $20x = 40$

19. $15 = 5y$

20. $65 = 13z$

21. $36 = 9a$

1. _____

2. _____

3. _____

4. _____

5. _____

6. _____

7. _____

8. _____

9. _____

10. _____

11. _____

12. _____

13. _____

14. _____

15. _____

16. _____

17. _____

18. _____

19. _____

20. _____

21. _____

Name _____   Score _____

Solve.

1. Eight times a number is forty. Find the number.

2. The product of fifteen and a number is equal to three hundred. Find the number.

1. _____

2. _____

3. The length of a rectangle is 7 inches more than the width. The length is 20 inches. Find the width of the rectangle.

4. Use the formula $d = rt$, where d is distance, r is the rate of speed and t is time, to find how long it would take to travel a distance of 1575 miles at a speed of 175 mph.

3. _____

4. _____

5. Use the formula $d = rt$, where d is distance, r is the rate of speed and t is time, to find how long it would take to travel a distance of 1290 miles at a speed of 215 mph.

6. Twenty-three added to a number equals fifty. Find the number.

5. _____

6. _____

7. The sum of twelve and a number equals forty-one. Find the number.

8. Use the formula $A = MN$, where A is the total amount paid M is the monthly payment, and N is the number of payments, to find the number of payments made of a loan for which the total amount paid is $18,960 and the monthly payment is $395.

7. _____

8. _____

9. Use the formula $A = MN$, where A is the total amount paid M is the monthly payment, and N is the number of payments, to find the number of payments made of a loan for which the total amount paid is $12,750 and the monthly payment is $425.

10. The length of a rectangle is 9 inches more than the width. The length is 17 inches. Find the width of the rectangle.

9. _____

10. _____

Name   Score

Simplify.

1. $8 - 2 + 4$ **2.** $9 - 3 + 5$ **3.** $7 + 4 \cdot 3$

4. $3 \cdot 3 - 9$ **5.** $10 - 4 \div 2$ **6.** $8 + 3 - 6$

7. $4^2 - 10$ **8.** $3(1 + 5) - 7$ **9.** $17 - 3^2$

10. $8 \cdot (7 + 3) \div 8$ **11.** $7 \cdot 3^2 + 16$ **12.** $2^2 - 2(12 \div 6)$

13. $13 + (9 - 7) \cdot 5$ **14.** $20 - 2 \cdot 3^2$ **15.** $4^2 + 6 \cdot (2 - 1)$

16. $2^4 + 7 \cdot (5 - 5)$ **17.** $3^2 \cdot 4 + 6 \cdot 2^3$ **18.** $8 \cdot 2 - 3^2$

19. $17 - 3 \cdot 5$ **20.** $14 + 4 \cdot 6$ **21.** $3 \cdot (8 - 7) + 10$

22. $11 - (11 - 2) \div 3$ **23.** $7 \cdot (8 - 5) + 12$ **24.** $14 \div (3 + 4) \cdot 2$

25. $8 - 5 + 9 \cdot 2 \div 3$ **26.** $6 \cdot 4 \div 3 \div 2 + 1$ **27.** $10(2 + 3) \div 5$

1. _____
2. _____
3. _____
4. _____
5. _____
6. _____
7. _____
8. _____
9. _____
10. _____
11. _____
12. _____
13. _____
14. _____
15. _____
16. _____
17. _____
18. _____
19. _____
20. _____
21. _____
22. _____
23. _____
24. _____
25. _____
26. _____
27. _____

Name   Score

Graph the numbers on the number line.

1.   4, −4

2.   0, 1

3.   −2, 1

4.   −3, −4

1.   __(see graph)__

2.   __(see graph)__

3.   __(see graph)__

4.   __(see graph)__

On the number line, which number is:

5.   7 units to the right of −5?

6.   3 units to the left of 2?

7.   8 units to the right of −4?

8.   6 units to the left of 2?

5.   _____

6.   _____

7.   _____

8.   _____

Place the correct symbol, < or >, between the two numbers.

9.   −7   0

10.   −4   −3

11.   5   −6

12.   −14   −24

13. $4\frac{1}{9}$ $-5\frac{2}{3}$

14. $-10\frac{1}{3}$ $-10\frac{1}{7}$

9.   _____

10.   _____

11.   _____

12.   _____

13.   _____

14.   _____

Write the given numbers from smallest to largest.

15.   6, −15, −8, 2

16.   9, −5, 0, 7

17.   3, 0, −9, 12

15.   _____

16.   _____

17.   _____

Name _____   Score _____

Find the opposite of the number.

1. 9	**2.** 3	**3.** −15	**1.** _____
			2. _____
			3. _____
4. 7	**5.** −4	**6.** −5	**4.** _____
			5. _____
			6. _____
7. 34	**8.** −28	**9.** 66	**7.** _____
			8. _____
			9. _____

Write the expression in words.

10. $-(-8)$	**11.** $1 + (-6)$	**12.** $-(-x)$	**10.** _____
			11. _____
			12. _____
13. $-15 - (-12)$	**14.** $-4 + (-3)$	**15.** $a - b$	**13.** _____
			14. _____
			15. _____

Simplify.

16. (-3)	**17.** $-(7)$	**18.** $-(-5)$	**16.** _____
			17. _____
			18. _____
19. $-(-13)$	**20.** $-(-4)$	**21.** 15	**19.** _____
			20. _____
			21. _____

17

Name   Score

Find the absolute value of the number.

1. $|-16|$ 2. $-|19|$ 3. $-|-20|$ 1. _____

                                                                                               2. _____

                                                                                               3. _____

Evaluate.

4. $-|65|$ 5. $|-10|$ 6. $-|-3|$ 4. _____

                                                                                               5. _____

                                                                                               6. _____

7. $|-0.6|$ 8. $\left|2\frac{6}{7}\right|$ 9. $-|-19|$ 7. _____

                                                                                               8. _____

                                                                                               9. _____

10. $|-28.1|$ 11. $-\left|-\frac{5}{8}\right|$ 12. $-|9.7|$ 10. _____

                                                                                               11. _____

                                                                                               12. _____

Place the correct symbol, <, = or > between the two numbers.

13. $|-14|$ $|17|$ 14. $-|4.03|$ $|-5|$ 15. $|35.4|$ $-|36|$ 13. _____

                                                                                               14. _____

                                                                                               15. _____

16. $-|1.5|$ $|-1.5|$ 17. $|-4|$ $|4|$ 18. $-|-9|$ $-|-15|$ 16. _____

                                                                                               17. _____

                                                                                               18. _____

Write the given numbers in order from smallest to largest.

19. $-|-1|,\ -3,\ |-4|,\ |6|$ 20. $|-10|,\ 7,\ |-3|,\ -|9|$ 21. $|2|,\ |-5|,\ -6,\ -|10|$ 19. _____

                                                                                               20. _____

                                                                                               21. _____

Name   Score
_____

Solve.

1. The daily high temperatures during one week were recorded as follows: –10°C, –4°C, 1°C, –11°C, –16°C, –7°C, and –2°C. What was the lowest daily high temperature for the week?

2. The daily high temperatures during one week were recorded as follows: –10°C, –4°C, 1°C, –11°C, –16°C, –7°C, and –2°C. What was the highest daily high temperature for the week?

1. _____

2. _____

3. The daily low temperatures during one week were recorded as follows: –20°C, –7°C, –18°C, –15°C, –23°C, –13°C, and –2°C. What was the highest daily low temperature for the week?

4. The daily low temperatures during one week were recorded as follows: –20°C, –7°C, –18°C, –15°C, –23°C, –13°C, and –2°C. What was the lowest daily low temperature for the week?

3. _____

4. _____

5. The value of a stock can increase and decrease every day. The daily increase or decrease for a stock during one week was recorded as follows: –1.05, –2.50, 0.74, –0.38, 0.49. What was the greatest increase for the stock for the week?

6. The value of a stock can increase and decrease every day. The daily increase or decrease for a stock during one week was recorded as follows: –6.27, 0.99, 0.75, –0.10, 0.53. What was the greatest decrease for the stock for the week?

5. _____

6. _____

7. The top four golfers in a tournament had scores –8, –5, 2, and –3 (under par). Which golfer had the best (smallest) score?

8. The top five golfers in a tournament had scores –11, 3, –4, –6 and –2 (under par). Which golfer had the worst (greatest) score?

7. _____

8. _____

9. A computer software company reported a loss of –35,500 in the first quarter and a loss of –37,400 in the second quarter. During which quarter was the loss greater?

10. A computer gaming company reported a loss of –13,400 in the third quarter and a loss of –11,800 in the fourth quarter. During which quarter was the loss greater?

9. _____

10. _____

Name   Score

Add.

1. $6 + 2$	**2.** $-5 + 3$	**3.** $-13 + 7$	**1.** _____
			2. _____
			3. _____
4. $44 + (-71)$	**5.** $-25 + (-16)$	**6.** $186 + (-98)$	**4.** _____
			5. _____
			6. _____
7. $4 + 9 + (-10)$	**8.** $-14 + (-3) + 8$	**9.** $13 + (-28) + 7$	**7.** _____
			8. _____
			9. _____
10. $-5 + 9 + (-18)$	**11.** $-7 + (-12) + 8$	**12.** $10 + (-4) + (-19)$	**10.** _____
			11. _____
			12. _____
13. $-12 + 37 + (-15)$	**14.** $41 + (-8) + (-16)$	**15.** $32 + (-23) + 11$	**13.** _____
			14. _____
			15. _____
16. $-6 + (-15) + (-9) + 18$	**17.** $4 + (-13) + 7 + (-17)$	**18.** $-21 + 30 + (-7) + (-12)$	**16.** _____
			17. _____
			18. _____

Evaluate the expression for the given values of the variables.

19. $x + y$, where $x = -30$ and $y = -22$

20. $-a + b - c$, where $a = 42$, $b = 28$ and $c = -14$

21. $x - y - z$, where $x = 29$, $y = -11$, $z = -4$

19. _____

20. _____

21. _____

Solve.

22. Is -2 a solution of the equation $x + 5 = 3$?

23. Is 5 a solution to the equation $8 - x = 13$?

24. Is -11 a solution to the equation $15 - a = 4$?

22. _____

23. _____

24. _____

Name _____   Score _____

Subtract.

1. $12 - 8$

2. $4 - 15$

3. $-5 - 16$

4. $42 - (-37)$

5. $-53 - (-37)$

6. $-108 - 95$

7. $-6 - 17 - (-11)$

8. $-16 - (-18) - (-13)$

9. $25 - 13 - (-7)$

10. $-42 - 23 - 9$

11. $79 - 12 - 53$

12. $-15 - (-33) - (-27)$

13. $-7 - (-15) - 13$

14. $8 - 23 - (-11)$

15. $-14 - (-3) - 9$

16. $-4 - 15 - (-28) - 5$

17. $23 - (-19) - 12 - 6$

18. $-5 - (-14) - (-9) - (-1)$

1. _____

2. _____

3. _____

4. _____

5. _____

6. _____

7. _____

8. _____

9. _____

10. _____

11. _____

12. _____

13. _____

14. _____

15. _____

16. _____

17. _____

18. _____

Evaluate the expression for the given values of the variables.

19. $-x - y$, where $x = -8$ and $y = -42$

20. $-a - b + c$, where $a = 3$, $b = -4$ and $c = -9$

21. $-x + y - z$, where $x = -12$, $y = -7$, $z = 3$

19. _____

20. _____

21. _____

Solve.

22. Is -8 a solution of the equation $x - 11 = -19$?

23. Is 9 a solution to the equation $4 = -b + 13$?

24. Is -3 a solution to the equation $6 - y = 3$?

22. _____

23. _____

24. _____

Name _____ Score _____

Solve.

1. Find the temperature after a rise of 9°C from –9°C.

2. Find the temperature after a rise of 4°C from –15°C.

1. _____

2. _____

3. During a game of Scrabble, Dan had a score of 70 points before his opponent used all the tiles, subtracting a score of 11 from Dan's total. What was Dan's score after his opponent ended the game?

4. During a game of Scrabble, Jean had a score of 115 points before she used all her tiles, entitling her to add 16 points to her score. What was Jean's score after she ended the game?

3. _____

4. _____

5. Find the temperature after a decrease of 10°F from 3°F.

6. Find the temperature after a decrease of 24°F from 52°F.

5. _____

6. _____

7. In a Minnesota city, the average nighttime temperature in January was –28°F and the average daytime temperature can reach 14°F. Find the difference between these average temperatures.

8. In a California city, the average nighttime temperature in July was 78°F and the average daytime temperature can reach 105°F. Find the difference between these average temperatures.

7. _____

8. _____

9. The elevation for Mt. Everest is 8850 meters and the elevation for the Dead Sea is –411 meters. What is the difference in elevation between Mt. Everest and the Dead Sea?

10. The elevation for Mt. McKinley is 5642 meters and the elevation for Death Valley is –28 meters. What is the difference in elevation between Mt. McKinley and Death Valley?

9. _____

10. _____

Name _____   Score _____

Multiply.

1. −4(8)   **2.** 8(−7)   **3.** −4(−9)   **1.** _____

                                                                                     2. _____

                                                                                     3. _____

4. −9·(−11)   **5.** 0·(−13)   **6.** −17(−29)   **4.** _____

                                                                                     5. _____

                                                                                     6. _____

7. −8·(−12)·(−4)   **8.** −16·4·10   **9.** 30·(−5)·0   **7.** _____

                                                                                     8. _____

                                                                                     9. _____

10. 6·(−13)   **11.** −13(−4)   **12.** −8(−15)   **10.** _____

                                                                                     11. _____

                                                                                     12. _____

13. 23(−5)   **14.** −6×12   **15.** 9(−14)   **13.** _____

                                                                                     14. _____

                                                                                     15. _____

Evaluate the expression for the given values of the variables.

16. $-xy$, where $x = -3$ **17.** $-abc$, where $a = -9$,   **18.** $-xyz$, where $x = 18$,   **16.** _____
and $y = 7$ $b = -7$ and $c = 8$ $y = 3$, $z = -6$

                                                                                     17. _____

                                                                                     18. _____

Solve.

19. Is −4 a solution of the   **20.** Is 0 a solution to the   **21.** Is −2 a solution to the   **19.** _____
equation $-2x = -8$?   equation $10 = -10y$?   equation $4a = -8$?

                                                                                     20. _____

                                                                                     21. _____

23

Name _____   Score _____

Divide.

1. $18 \div (-6)$ **2.** $-54 \div 6$ **3.** $-35 \div (-5)$

4. $35 \div (-7)$ **5.** $-96 \div (-8)$ **6.** $-54 \div 0$

7. $141 \div (-3)$ **8.** $0 \div (-30)$ **9.** $-174 \div (-6)$

10. $84 \div (-7)$ **11.** $-924 \div 11$ **12.** $-280 \div (-4)$

13. $-56 \div 4$ **14.** $-32 \div (-8)$ **15.** $910 \div (-13)$

Evaluate the expression for the given values of the variables.

16. $x \div y$, where $x = -75$ and $y = -5$ **17.** $a \div (-b)$, where $a = 72$ and $b = 9$ **18.** $-x \div y$, where $x = -42$ and $y = 6$

Solve.

19. Is 0 a solution of the equation $\frac{0}{-12} = x$?   **20.** Is -4 a solution to the equation $\frac{a}{24} = 8$?   **21.** Is 3 a solution to the equation $\frac{-15}{y} = -5$?

1. _____
2. _____
3. _____
4. _____
5. _____
6. _____
7. _____
8. _____
9. _____
10. _____
11. _____
12. _____
13. _____
14. _____
15. _____
16. _____
17. _____
18. _____
19. _____
20. _____
21. _____

Name   Score

Solve.

1. Find the temperature after a rise of 4°C from –2°C.

2. Find the temperature after a rise of 6°C from –10°C.

1. _____

2. _____

3. During a card game of Hearts, you had a score of 14 points before your opponent "shot the moon," subtracting a score of 26 from your total. What was your score after your opponent "shot the moon"?

4. In a card game of Hearts, you had a score of –15 before you "shot the moon," entitling you to add 26 points to your score. What was your score after you "shot the moon"?

3. _____

4. _____

5. The daily low temperatures during one week were recorded as follows: 4°, –6°, 8°, –2°, –9°, –11°, and –5°. Find the average daily low temperature for the week.

6. The daily high temperatures during one week were recorded as follows: –5°, –8°, 6°, 8°, 0°, –6°, and –2°. Find the average daily high temperature for the week.

5. _____

6. _____

7. One golfer had a score of six under par (–6) while a second golfer had a score of 8 over par (+8). Find the difference between their scores.

8. A four-member golf team had a combined score of 18 under par (–18). Another team had a combined score of 2 over par (+2). Find the difference between their scores.

7. _____

8. _____

9. During the last week in June a stock rose 15 points. The stock then fell 9 points during the next week. Find the net change in the value of the stock for the two week period.

10. During the first week in May a stock fell 11 points. Then the stock rose 6 points during the next week. Find the net change in the value of the stock for the two week period.

9. _____

10. _____

Name   Score

Solve.

1. $x - 5 = 7$

2. $m - 3 = 7$

3. $9 = y - 5$

4. $14 = t - 6$

5. $x - 6 = -13$

6. $n - 9 = -25$

7. $-12 = z + 4$

8. $n - 8 = -8$

9. $-13 = c - 13$

10. $5 + x = 3$

11. $15 = b + 22$

12. $5m = -15$

13. $7p = -42$

14. $-22 = 11v$

15. $-5v = 60$

16. $-6y = -24$

17. $4x = -52$

18. $5x = 0$

19. $-6n = 0$

20. $-17 = -17z$

21. $4x = -80$

22. $-45 = -15v$

1. _____

2. _____

3. _____

4. _____

5. _____

6. _____

7. _____

8. _____

9. _____

10. _____

11. _____

12. _____

13. _____

14. _____

15. _____

16. _____

17. _____

18. _____

19. _____

20. _____

21. _____

22. _____

Name _____   Score _____

Solve.

1. Eleven less than a number is
    twenty-six. Find the number.

2. Zero is equal to nineteen more than
    some number. Find the number.

1. _____

2. _____

3. Twenty-five equals the sum of a number
    and forty-one. Find the number.

4. Eighteen equals negative three times a
    number. Find the number.

3. _____

4. _____

5. The temperature now is $7°$ higher than
    it was this morning. The temperature
    now is $11°C$. What was the temperature
    this morning?

6. An office supplier wants to make a
    profit of $115 on the sale of a software
    package that cost the supplier $435.
    Use the equation $P = S - C$, where P
    is the profit on an item, S is the selling
    price, and C is the cost, to find the
    selling price of the software.

5. _____

6. _____

7. The net worth of a business is given
    by the formula $N - A - L$, where N is
    the net worth, A is the assets of the
    business (or the amount owned), and
    L is the liabilities of the business (or
    the amount owed). Use this formula
    to find the assets of a business that
    has a net worth of $12 million and
    liabilities of $3 million.

8. The net worth of a company is $93
    million and it has liabilities of $24 million.
    Use the net worth formula $N - A - L$,
    where N is the net worth, A is the
    assets of the business (or the amount
    owned), and L is the liabilities of the
    business (or the amount owed), to find
    the assets of the company.

7. _____

8. _____

Name   Score

Simplify.

1. $9 \div 3 + 7$ 2. $2 - 15 \div 5$ 1. _____

                                                                                  2. _____

3. $7 - (3^2) \times 5$ 4. $14 \div 7 - 6 \div 2$ 3. _____

                                                                                  4. _____

5. $(-5)^2 \times 2 \div (9+1)$ 6. $9 \times 2 - 4 \times 5 + 6 \times 3 + 7 - 3$ 5. _____

                                                                                  6. _____

7. $-5 \times (-3)^2 \times 2 \div 3 - (-10)$ 8. $2^2 \times (3-4) \div 2 + 5 - 8 \times 2$ 7. _____

                                                                                  8. _____

9. $-6 \cdot 8 \div (-3)$ 10. $2 \cdot (-4) - 5$ 9. _____

                                                                                  10. _____

11. $(7-3) \cdot (-4)$ 12. $(-3)^2 \div 3$ 11. _____

                                                                                  12. _____

Evaluate the variable expression given $a = -3$, $b = -4$, $c = 0$, and $d = 2$.

13. $4a - 3c$ 14. $\dfrac{d-b}{a}$ 13. _____

                                                                                  14. _____

15. $\dfrac{3d - 2a}{4c}$ 16. $(d-a)^2 + (c-b)^2$ 15. _____

                                                                                  16. _____

Name   Score

Find the LCM.

1. 3, 4   2. 3, 7   3. 6, 9   1. _____

                                                                     2. _____

                                                                     3. _____

4. 8, 10   5. 4, 8   6. 9, 12   4. _____

                                                                      5. _____

                                                                     6. _____

7. 4, 9   8. 6, 15   9. 16, 24   7. _____

                                                                      8. _____

                                                                     9. _____

10. 15, 25   11. 28, 32   12. 4, 18   10. _____

                                                                      11. _____

                                                                     12. _____

13. 72, 108   14. 84, 126   15. 32, 128   13. _____

                                                                      14. _____

                                                                     15. _____

16. 3, 7, 9   17. 6, 12, 27   18. 3, 7, 11   16. _____

                                                                      17. _____

                                                                     18. _____

19. 9, 12, 24   20. 10, 25, 40   21. 4, 7, 21   19. _____

                                                                      20. _____

                                                                     21. _____

22. 2, 7, 11   23. 28, 32, 56   24. 16, 20, 40   22. _____

                                                                      23. _____

                                                                     24. _____

25. 8, 27, 36   26. 6, 12, 18   27. 2, 16, 32   25. _____

                                                                      26. _____

                                                                      27. _____

Name   Score

Find the GCF.

1. 3, 7	2. 3, 6	3. 9, 16
4. 8, 18	5. 10, 15	6. 30, 65
7. 15, 30	8. 36, 56	9. 18, 27
10. 21, 35	11. 15, 20	12. 30, 50
13. 48, 64	14. 39, 52	15. 37, 67
16. 4, 8, 10	17. 3, 5, 7	18. 3, 9, 12
19. 5, 11, 13	20. 10, 25, 30	21. 16, 40, 80
22. 16, 20, 32	23. 24, 32, 40	24. 18, 27, 81
25. 28, 44, 56	26. 17, 68, 85	27. 30, 75, 150

1. _____
2. _____
3. _____
4. _____
5. _____
6. _____
7. _____
8. _____
9. _____
10. _____
11. _____
12. _____
13. _____
14. _____
15. _____
16. _____
17. _____
18. _____
19. _____
20. _____
21. _____
22. _____
23. _____
24. _____
25. _____
26. _____
27. _____

Name _____   Score _____

Read the given exercise and state whether you will use an LCM or GCF to solve the problem.

1. Exercise 5   2. Exercise 6   1. _____

                                                                   2. _____

3. Exercise 7   4. Exercise 8   3. _____

                                                                   4. _____

Solve.

5. A discount catalog offers socks   6. Each week, copies of a national   5. _____
    at reduced prices. The customer   magazine are delivered to three
    must order 4 pairs, 8 pairs or 16   different stores that have ordered
    pairs of socks. How many pairs   60 copies, 80 copies, and 120 copies,
    should be packaged together so   respectively. How many copies
    that no package needs to be opened   should be packages together so that
    when a clerk is filling an order?   no package needs to be opened   6. _____
                                               during delivery?

7. Two machines are filling boxes of   8. You and a friend are swimming laps   7. _____
    dried pasta. One machine, which is   at a pool. You swim one lap every 5
    filling 16-oz boxes, fills one box   minutes. Your friend swims one lap
    every 3 minutes. The second machine,   every 6 minutes. If you start at the
    which is filling 24-oz boxes, fills one   same time from the same end of the
    box every 4 minutes. How often are   pool, in how many minutes will both
    the two machines starting to fill a box   of you be at the starting point again? 8. _____
    at the same time?

Name   Score

Express the shaded portion of the circle as a fraction.

1.   2.

1. _____

2. _____

Express the shaded portion of the circles as a mixed number.

3.    4.

5.   6.

3. _____

4. _____

5. _____

6. _____

Express the shaded portion of the circles as an improper fraction.

7.   8.

7. _____

8. _____

Write the improper fraction as a mixed number or whole number.

9. $\dfrac{11}{3}$ 10. $\dfrac{30}{5}$ 11. $\dfrac{17}{9}$

9. _____

10. _____

11. _____

12. $\dfrac{36}{12}$ 13. $\dfrac{11}{6}$ 14. $\dfrac{8}{8}$

12. _____

13. _____

14. _____

Write the mixed number as an improper fraction.

15. $2\dfrac{1}{5}$ 16. $7\dfrac{2}{3}$ 17. $4\dfrac{7}{9}$

15. _____

16. _____

17. _____

18. $10\dfrac{3}{4}$ 19. $9\dfrac{7}{10}$ 20. $12\dfrac{5}{6}$

18. _____

19. _____

20. _____

Name   Score

Build an equivalent fraction with the given denominator.

1. $\dfrac{2}{3} = \dfrac{}{39}$

2. $\dfrac{5}{6} = \dfrac{}{18}$

3. $\dfrac{4}{7} = \dfrac{}{56}$

4. $\dfrac{1}{5} = \dfrac{}{25}$

5. $\dfrac{3}{4} = \dfrac{}{28}$

6. $\dfrac{6}{10} = \dfrac{}{30}$

7. $\dfrac{1}{2} = \dfrac{}{30}$

8. $\dfrac{4}{5} = \dfrac{}{45}$

9. $\dfrac{2}{9} = \dfrac{}{54}$

10. $\dfrac{6}{7} = \dfrac{}{84}$

11. $6 = \dfrac{}{9}$

12. $8 = \dfrac{}{15}$

1. _____

2. _____

3. _____

4. _____

5. _____

6. _____

7. _____

8. _____

9. _____

10. _____

11. _____

12. _____

Reduce the fraction to simplest form.

13. $\dfrac{8}{12}$

14. $\dfrac{15}{25}$

15. $\dfrac{2}{16}$

16. $\dfrac{28}{49}$

17. $\dfrac{54}{81}$

18. $\dfrac{40}{64}$

19. $\dfrac{60}{96}$

20. $\dfrac{18}{36}$

21. $\dfrac{7}{18}$

22. $\dfrac{25}{100}$

23. $\dfrac{84}{144}$

24. $\dfrac{39}{13}$

13. _____

14. _____

15. _____

16. _____

17. _____

18. _____

19. _____

20. _____

21. _____

22. _____

23. _____

24. _____

Name   Score

Place the correct symbol, < or >, between the two numbers.

1. $\frac{11}{17}$ $\frac{14}{17}$ 2. $\frac{4}{5}$ $\frac{5}{9}$ 3. $\frac{7}{10}$ $\frac{2}{3}$

4. $\frac{7}{12}$ $\frac{11}{15}$ 5. $\frac{5}{7}$ $\frac{3}{4}$ 6. $\frac{2}{3}$ $\frac{3}{5}$

7. $\frac{3}{10}$ $\frac{1}{6}$ 8. $\frac{7}{9}$ $\frac{9}{14}$ 9. $\frac{4}{11}$ $\frac{1}{2}$

10. $\frac{1}{4}$ $\frac{4}{15}$ 11. $\frac{13}{20}$ $\frac{5}{7}$ 12. $\frac{8}{13}$ $\frac{11}{18}$

13. $\frac{19}{24}$ $\frac{25}{36}$ 14. $\frac{8}{15}$ $\frac{23}{35}$ 15. $\frac{5}{9}$ $\frac{3}{5}$

16. $\frac{5}{22}$ $\frac{3}{8}$ 17. $\frac{1}{4}$ $\frac{5}{26}$ 18. $\frac{5}{12}$ $\frac{4}{9}$

19. $\frac{9}{16}$ $\frac{10}{17}$ 20. $\frac{19}{22}$ $\frac{39}{46}$ 21. $\frac{23}{30}$ $\frac{17}{20}$

22. $\frac{5}{6}$ $\frac{3}{4}$ 23. $\frac{11}{14}$ $\frac{15}{19}$ 24. $\frac{7}{24}$ $\frac{8}{15}$

25. $\frac{4}{5}$ $\frac{16}{21}$ 26. $\frac{7}{15}$ $\frac{5}{8}$ 27. $\frac{21}{25}$ $\frac{27}{35}$

1. _____
2. _____
3. _____
4. _____
5. _____
6. _____
7. _____
8. _____
9. _____
10. _____
11. _____
12. _____
13. _____
14. _____
15. _____
16. _____
17. _____
18. _____
19. _____
20. _____
21. _____
22. _____
23. _____
24. _____
25. _____
26. _____
27. _____

Name _____   Score _____

Solve.

1. A class has 40 students enrolled. If 35 students attend the class, what fractional part of the class is present?

2. A gallon is equal to 64 oz. What fractional part of a gallon is 24 oz?

1. _____

2. _____

3. If you work for 6 h one day, what fractional part of one day did you spend working?

4. You spend 40 minutes preparing a meal. What fractional part of an hour did you spend making preparations?

3. _____

4. _____

5. A standard deck of playing cards consists of 52 cards. What fractional part of a standard deck of cards is kings?

6. Gold is designated by karats. Pure gold is 24 karats. What fraction part of a 16-karat gold necklace is pure gold?

5. _____

6. _____

7. To pass a drivers license examination you must answer at least $\frac{7}{10}$ of the question correctly. If the exam has 20 questions and you answer 16 correctly, do you pass the exam?

8. You answer 44 questions correctly on an exam of 50 questions. Did you answer more or less than $\frac{9}{10}$ of the questions correctly?

7. _____

8. _____

Name   Score

Multiply.

1. $\dfrac{4}{5}\cdot\dfrac{10}{13}$

2. $\dfrac{3}{4}\cdot\dfrac{8}{11}$

3. $\dfrac{4}{15}\cdot\dfrac{5}{9}$

4. $\dfrac{14}{15}\cdot\dfrac{7}{12}$

5. $\dfrac{5}{2}\cdot\dfrac{23}{20}$

6. $\dfrac{8}{15}\cdot\dfrac{6}{7}$

7. $\dfrac{9}{10}\cdot\dfrac{13}{18}$

8. $\dfrac{3}{7}\cdot\dfrac{5}{8}$

9. $\dfrac{12}{19}\cdot\dfrac{19}{6}$

10. $\dfrac{2}{3}\cdot 3\dfrac{3}{4}$

11. $\dfrac{5}{6}\cdot 30$

12. $1\dfrac{4}{9}\cdot\dfrac{3}{10}$

13. $\dfrac{4}{7}\cdot 2\dfrac{7}{12}$

14. $0\cdot 3\dfrac{1}{6}$

15. $10\cdot 5\dfrac{8}{15}$

16. $8\dfrac{3}{4}\cdot 6\dfrac{4}{5}$

17. $3\dfrac{5}{24}\cdot 12$

18. $2\dfrac{1}{7}\cdot 16\dfrac{4}{5}$

Evaluate the variable expression xy for the given values of x and y.

19. $x=\dfrac{7}{12}\,;\,y=\dfrac{16}{9}$

20. $x=\dfrac{3}{8}\,;\,y=\dfrac{10}{17}$

21. $x=\dfrac{6}{11}\,;\,y=\dfrac{1}{4}$

1. ___ 2. ___ 3. ___ 4. ___ 5. ___ 6. ___ 7. ___ 8. ___ 9. ___ 10. ___ 11. ___ 12. ___ 13. ___ 14. ___ 15. ___ 16. ___ 17. ___ 18. ___ 19. ___ 20. ___ 21. ___

Name _____   Score _____

Divide.

1. $\dfrac{4}{5} \div \dfrac{8}{5}$

2. $\dfrac{5}{24} \div \dfrac{5}{12}$

3. $\dfrac{10}{19} \div \dfrac{14}{19}$

4. $0 \div \dfrac{1}{4}$

5. $\dfrac{3}{15} \div \dfrac{8}{55}$

6. $\dfrac{2}{15} \div \dfrac{7}{12}$

7. $\dfrac{9}{20} \div \dfrac{11}{30}$

8. $\dfrac{14}{15} \div 7$

9. $\dfrac{5}{18} \div \dfrac{5}{9}$

10. $\dfrac{1}{2} \div \dfrac{9}{8}$

11. $\dfrac{3}{8} \div 0$

12. $\dfrac{3}{7} \div \dfrac{9}{14}$

13. $4 \div 2\dfrac{2}{3}$

14. $\dfrac{4}{5} \div 2\dfrac{3}{4}$

15. $3\dfrac{3}{5} \div 9$

16. $1\dfrac{7}{12} \div \dfrac{7}{8}$

17. $21\dfrac{1}{4} \div 17$

18. $4\dfrac{3}{4} \div 3\dfrac{1}{3}$

19. $7\dfrac{4}{8} \div \dfrac{4}{15}$

20. $18\dfrac{6}{22} \div 7\dfrac{1}{11}$

21. $20\dfrac{5}{9} \div 6\dfrac{2}{3}$

1. _____
2. _____
3. _____
4. _____
5. _____
6. _____
7. _____
8. _____
9. _____
10. _____
11. _____
12. _____
13. _____
14. _____
15. _____
16. _____
17. _____
18. _____
19. _____
20. _____
21. _____

Evaluate the variable expression $x \div y$ for the given values of x and y.

22. $x = \dfrac{3}{4}; y = \dfrac{15}{6}$

23. $x = \dfrac{4}{11}; y = \dfrac{12}{17}$

24. $x = 3; y = \dfrac{9}{11}$

22. _____
23. _____
24. _____

Name   Score

Solve.

1. An apprentice bricklayer earns $24 an hour. What are the bricklayer's total earnings after working $7\frac{1}{4}$ hours?

1. _____

2. A sports car gets 21 miles on each gallon of gasoline. How many miles can the car travel on $5\frac{2}{3}$ gallons of gasoline?

2. _____

3. A station wagon used $15\frac{3}{10}$ gallons of gasoline on a 306-mile trip. How many miles did this car travel on one gallon of gasoline?

3. _____

4. A volunteer for the psychology experiment answered "yes" to 18 of the questions on one questionnaire. This was $\frac{3}{8}$ of the total number of questions. Find the total number of questions on the questionnaire.

4. _____

5. A board is $8\frac{3}{4}$ feet long. One fourth of the board is cut off. What is the length of the piece cut off?

5. _____

6. A plumber earns $320 for each day worked. What is the total of the plumber's earnings for working $4\frac{2}{5}$ days.

6. _____

7. If 55,000 people voted for the new city ordinance, and this was $\frac{5}{6}$ of the total number of registered voters in the city, how many registered voters are there?

7. _____

8. A $2\frac{1}{2}$-grain precious metal ingot sold for $30. Find the price per grain of the precious metal.

8. _____

9. A student read $\frac{2}{5}$ of a book containing 645 pages. How many pages did the student read?

9. _____

10. A person can walk $3\frac{1}{4}$ miles in one hour. How many miles can the person walk in $1\frac{3}{4}$ hours?

10. _____

11. An investor purchased a $3\frac{1}{4}$-ounce gold coin for $2028. What was the price for one ounce?

11. _____

12. A car traveled 143 miles in $3\frac{1}{4}$ hours. What was the car's average speed in miles per hour?

12. _____

Name   Score

Add.

1. $\dfrac{3}{6}+\dfrac{2}{6}$

2. $\dfrac{5}{10}+\dfrac{6}{10}$

3. $\dfrac{7}{19}+\dfrac{3}{19}$

4. $\dfrac{3}{15}+\dfrac{9}{15}+\dfrac{1}{15}$

5. $\dfrac{6}{10}+\dfrac{1}{10}+\dfrac{8}{10}$

6. $\dfrac{3}{4}+\dfrac{6}{4}+\dfrac{8}{4}$

7. $\dfrac{1}{4}+\dfrac{5}{6}$

8. $\dfrac{4}{15}+\dfrac{5}{9}$

9. $\dfrac{2}{3}+\dfrac{7}{8}$

10. $\dfrac{13}{14}+\dfrac{15}{18}$

11. $\dfrac{3}{14}+\dfrac{17}{21}$

12. $\dfrac{9}{10}+\dfrac{10}{16}$

13. $\dfrac{1}{4}+\dfrac{1}{3}+\dfrac{8}{9}$

14. $\dfrac{1}{2}+\dfrac{5}{8}+\dfrac{10}{12}$

15. $\dfrac{3}{5}+\dfrac{6}{7}+\dfrac{5}{9}$

16. $\dfrac{9}{12}+\dfrac{2}{4}+\dfrac{5}{16}$

17. $\dfrac{2}{3}+\dfrac{7}{9}+\dfrac{7}{10}$

18. $\dfrac{4}{5}+\dfrac{1}{6}+\dfrac{15}{18}$

19. $\dfrac{3}{6}+\dfrac{1}{8}+\dfrac{7}{10}$

20. $\dfrac{1}{5}+\dfrac{1}{11}+\dfrac{1}{15}$

21. $\dfrac{5}{6}+\dfrac{15}{24}+\dfrac{7}{8}$

22. $6\dfrac{1}{2}+5\dfrac{2}{3}$

23. $7\dfrac{5}{6}+2\dfrac{13}{15}$

24. $4\dfrac{3}{5}+6\dfrac{4}{7}$

25. $3\dfrac{1}{3}+2\dfrac{1}{2}+9\dfrac{3}{4}$

26. $7\dfrac{4}{7}+1\dfrac{3}{14}+11\dfrac{2}{5}$

27. $6\dfrac{7}{9}+5\dfrac{11}{12}+10\dfrac{5}{10}$

1. _____
2. _____
3. _____
4. _____
5. _____
6. _____
7. _____
8. _____
9. _____
10. _____
11. _____
12. _____
13. _____
14. _____
15. _____
16. _____
17. _____
18. _____
19. _____
20. _____
21. _____
22. _____
23. _____
24. _____
25. _____
26. _____
27. _____

Name _____ Score _____

Subtract.

1. $\dfrac{7}{12} - \dfrac{5}{12}$

2. $\dfrac{17}{24} - \dfrac{7}{24}$

3. $\dfrac{13}{18} - \dfrac{7}{18}$

4. $\dfrac{43}{56} - \dfrac{19}{56}$

5. $\dfrac{24}{27} - \dfrac{10}{27}$

6. $\dfrac{5}{13} - \dfrac{3}{13}$

7. $\dfrac{22}{27} - \dfrac{14}{45}$

8. $\dfrac{33}{40} - \dfrac{7}{16}$

9. $\dfrac{1}{6} - \dfrac{2}{13}$

10. $\dfrac{14}{15} - \dfrac{7}{12}$

11. $\dfrac{6}{7} - \dfrac{3}{5}$

12. $\dfrac{9}{14} - \dfrac{12}{35}$

13. $\dfrac{17}{24} - \dfrac{5}{9}$

14. $\dfrac{23}{26} - \dfrac{34}{39}$

15. $\dfrac{5}{8} - \dfrac{9}{20}$

16. $15\dfrac{15}{26} - 6$

17. $12 - 7\dfrac{15}{16}$

18. $17\dfrac{2}{5} - 9\dfrac{4}{5}$

19. $44 - 25\dfrac{4}{7}$

20. $36\dfrac{5}{14} - 14\dfrac{9}{14}$

21. $27\dfrac{2}{5} - 23\dfrac{4}{5}$

22. $33\dfrac{8}{15} - 19\dfrac{8}{9}$

23. $62\dfrac{17}{24} - 28\dfrac{11}{16}$

24. $69\dfrac{19}{25} - 60\dfrac{23}{25}$

25. $43\dfrac{16}{21} - 34\dfrac{25}{42}$

26. $12\dfrac{2}{5} - 9\dfrac{5}{6}$

27. $289\dfrac{9}{14} - 163\dfrac{3}{8}$

1. _____
2. _____
3. _____
4. _____
5. _____
6. _____
7. _____
8. _____
9. _____
10. _____
11. _____
12. _____
13. _____
14. _____
15. _____
16. _____
17. _____
18. _____
19. _____
20. _____
21. _____
22. _____
23. _____
24. _____
25. _____
26. _____
27. _____

Name   Score

Solve.

1. Over a two-year period, a child grew $4\frac{5}{8}$ inches. If the child grew $1\frac{1}{2}$ inches the first year, how many inches did the child grow during the second year?

2. A wall that is $\frac{3}{4}$-inches thick is covered by a $\frac{5}{16}$-inch veneer. Find the total thickness after the veneer is installed.

1. _____

2. _____

3. Find the missing dimension.

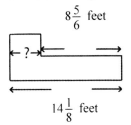

$8\frac{5}{6}$ feet

? →

$14\frac{1}{8}$ feet

4. For an upcoming role, an actor is put on a diet to lose 15 pounds. If $3\frac{1}{2}$ pounds are lost the first week and $2\frac{1}{4}$ pounds the second week, how many pounds must be lost to achieve the goal?

3. _____

4. _____

5. A carpenter build a header by nailing a $1\frac{1}{4}$-inch board to a $2\frac{5}{8}$-inch beam. Find the total thickness of the header.

6. You bought stock in a publishing company for $18\frac{5}{8}$ ($\$18\frac{5}{8}$ per share). The price of the stock gained $4\frac{1}{2}$ ($\$4\frac{1}{2}$ per share) during a three-week period. Find the price of the stock at the end of the three weeks.

5. _____

6. _____

7. An artist places a $2\frac{7}{8}$-inch mat around a painting that is $16\frac{1}{2}$ inches long and $11\frac{3}{4}$ inches wide. Find the new dimensions of the picture.

8. A shopper bought $\frac{1}{2}$ pound of cheese, $\frac{3}{4}$ pound of ham, $\frac{3}{4}$ pound of bologna, and $1\frac{1}{4}$ pounds of salami. How many pounds of cold cuts did the shopper buy?

7. _____

8. _____

9. A planer removes $\frac{1}{4}$ inch from a $\frac{5}{6}$-inch board. Find the resulting thickness of the board.

10. A $2\frac{2}{9}$-foot piece is cut from a 4-foot board. How much of the board is left?

9. _____

10. _____

Name   Score

Solve.

1. $\dfrac{x}{5} = 8$

2. $-5 = \dfrac{m}{7}$

3. $\dfrac{4}{5}x = 20$

4. $x + \dfrac{1}{3} = \dfrac{6}{7}$

5. $\dfrac{8}{9} = y - \dfrac{1}{7}$

6. $-\dfrac{6a}{7} = \dfrac{3}{4}$

7. $15a = 25$

8. $-8c = 14$

9. $14z = 21$

10. $-6y = -\dfrac{18}{25}$

11. $\dfrac{5z}{6} = -\dfrac{7}{12}$

12. $-4x = \dfrac{12}{7}$

13. $-\dfrac{3}{7}t = -\dfrac{11}{14}$

14. $\dfrac{9}{11} = y + \dfrac{4}{5}$

15. $-\dfrac{1}{3}x = 4$

16. $-12c = 15$

17. $-6x = \dfrac{3}{7}$

18. $-3y = -\dfrac{12}{13}$

19. $-11z = 55$

20. $-9z = 24$

21. $-6y = -\dfrac{12}{19}$

1. _____

2. _____

3. _____

4. _____

5. _____

6. _____

7. _____

8. _____

9. _____

10. _____

11. _____

12. _____

13. _____

14. _____

15. _____

16. _____

17. _____

18. _____

19. _____

20. _____

21. _____

Name _____   Score _____

Solve.

1. A number minus one-fifth equals one-third. Find the number.

2. Four-fifths times a number is eight-ninths. Find the number.

1. _____

2. _____

3. The quotient of a number and negative six is five-sixths. Find the number.

4. Negative three-sevenths equals the product of four-fifths and some number. Find the number.

3. _____

4. _____

5. The number of liters of ginger ale in a sparkling punch is two-fifths the total number of liters in the punch. The number of liters of ginger ale is 10. Find the total number of liters in the punch.

6. The number of people who voted in an election for mayor of a city was three-fourths of the total number of eligible voters. There were 42,588 people who voted in the election. Find the number of eligible voters.

5. _____

6. _____

7. The average number of miles per gallon for a car is calculated using the formula $a = \frac{m}{g}$, where a is the average number of miles per gallon and m is the number of miles traveled on g gallons of gas. Use this formula to find the number of miles a car can travel on 17 gal of gas if the car averages 32 miles per gallon.

8. The average number of miles per gallon for a truck is calculated using the formula $a = \frac{m}{g}$, where a is the average number of miles per gallon and m is the number of miles traveled on g gallons of gas. Use this formula to find the number of miles a truck can travel on 37 gal of diesel fuel if the truck averages 13 miles per gallon.

7. _____

8. _____

Name   Score

Evaluate.

1. $\left(\dfrac{5}{11}\right)^2$ 2. $\left(\dfrac{6}{7}\right)^2$ 3. $\left(-\dfrac{1}{8}\right)^2$ 1. _____

                                                                           2. _____

                                                                           3. _____

4. $\left(-\dfrac{2}{3}\right)^3$ 5. $\left(3\dfrac{1}{3}\right)^2$ 6. $\left(1\dfrac{1}{5}\right)^2$ 4. _____

                                                                           5. _____

                                                                           6. _____

7. $\left(\dfrac{3}{5}\right)^2\left(\dfrac{5}{7}\right)$ 8. $\left(\dfrac{1}{6}\right)^2\left(\dfrac{6}{7}\right)^2$ 9. $\left(-\dfrac{1}{2}\right)^3\left(\dfrac{2}{3}\right)^2$ 7. _____

                                                                           8. _____

                                                                           9. _____

10. $\left(\dfrac{7}{10}\right)^2\left(\dfrac{5}{6}\right)$ 11. $3^2\cdot\left(-\dfrac{5}{6}\right)^2$ 12. $\left(\dfrac{2}{3}\right)^2\left(\dfrac{1}{2}\right)^3\left(-\dfrac{6}{5}\right)$ 10. _____

                                                                          11. _____

                                                                          12. _____

Evaluate the variable expression for the given values of x and y.

13. x^4 14. x^3y^4 15. x^3y^2 13. _____

    for $x=\dfrac{2}{5}$ for $x=\dfrac{5}{6}$ and $y=\dfrac{6}{10}$ for $x=1\dfrac{1}{3}$ and $y=\dfrac{3}{4}$

                                                                          14. _____

                                                                          15. _____

Name   Score

Simplify.

1. $\dfrac{\frac{3}{10}}{\frac{4}{5}}$

2. $\dfrac{\frac{11}{30}}{\frac{5}{6}}$

1. _____

2. _____

3. $\dfrac{4-\frac{5}{8}}{\frac{3}{16}}$

4. $\dfrac{\frac{1}{3}}{\frac{4}{5}-\frac{3}{10}}$

3. _____

4. _____

5. $\dfrac{\frac{4}{9}}{\frac{5}{6}+\frac{2}{3}}$

6. $\dfrac{\frac{1}{4}+\frac{3}{8}}{\frac{5}{8}}$

5. _____

6. _____

7. $\dfrac{\frac{2}{7}-\frac{3}{21}}{\frac{3}{7}-\frac{5}{21}}$

8. $\dfrac{\frac{5}{9}+\frac{2}{27}}{\frac{4}{9}+\frac{5}{27}}$

7. _____

8. _____

9. $\dfrac{\frac{7}{8}-\frac{1}{4}}{\frac{1}{3}+\frac{2}{9}}$

10. $\dfrac{\frac{3}{4}+\frac{3}{16}}{\frac{7}{8}-\frac{5}{16}}$

9. _____

10. _____

11. $\dfrac{1\frac{2}{3}+2\frac{1}{3}}{4\frac{2}{9}-3\frac{5}{9}}$

12. $\dfrac{3\frac{1}{4}-2\frac{1}{2}}{5\frac{1}{3}+\frac{2}{3}}$

11. _____

12. _____

Evaluate the expression for the given values of the variables.

13. $\dfrac{x+y}{z}$, when $x=\frac{3}{8}$, $y=\frac{2}{3}$, and $z=\frac{1}{12}$

14. $\dfrac{x}{y-z}$, when $x=\frac{5}{8}$, $y=\frac{1}{2}$, and $z=\frac{2}{5}$

13. _____

14. _____

Name _____   Score _____

Simplify using the Order of Operations Agreement.

1. $\dfrac{1}{4} + \dfrac{3}{7} - \dfrac{1}{6}$

2. $\dfrac{2}{3} - \dfrac{1}{5} + \dfrac{4}{9}$

3. $\dfrac{5}{8} + \dfrac{2}{9} \div \dfrac{8}{9}$

4. $\dfrac{4}{7} \div \dfrac{6}{11} + \dfrac{4}{5}$

5. $\dfrac{3}{4} \cdot \dfrac{5}{12} + \dfrac{9}{16}$

6. $\dfrac{1}{2} + \dfrac{5}{8} \cdot \dfrac{3}{10}$

7. $\left(\dfrac{5}{6}\right)^2 - \dfrac{2}{7}$

8. $\dfrac{7}{10} - \left(\dfrac{2}{3}\right)^3$

9. $\dfrac{5}{12} \cdot \left(\dfrac{4}{5} - \dfrac{8}{15}\right) + \dfrac{1}{3}$

10. $\dfrac{7}{8} \div \left(\dfrac{1}{4} + \dfrac{1}{10}\right) - \dfrac{7}{11}$

11. $\dfrac{3}{8} - \left(\dfrac{1}{2}\right)^3 + \dfrac{4}{9}$

12. $\dfrac{2}{5} + \left(\dfrac{2}{5}\right)^2 - \dfrac{4}{25}$

13. $\dfrac{1}{8} \div \left(\dfrac{1}{2}\right)^2 - \dfrac{1}{2}$

14. $\left(\dfrac{4}{7}\right)^2 \cdot \dfrac{3}{4} + \dfrac{11}{14}$

15. $\left(\dfrac{5}{8} + \dfrac{1}{4}\right) \cdot \dfrac{16}{21}$

16. $\dfrac{5}{6} \div \left(\dfrac{9}{10} - \dfrac{2}{15}\right)$

17. $\left(\dfrac{5}{9} + \dfrac{13}{18}\right) \div \left(\dfrac{2}{3}\right)^2$

18. $\left(\dfrac{1}{4}\right)^2 \cdot \left(\dfrac{7}{9} - \dfrac{1}{3}\right)$

19. $\dfrac{6}{7} \cdot \dfrac{14}{27} \div \dfrac{4}{9}$

20. $\left(\dfrac{1}{3}\right)^2 + \left(\dfrac{4}{5} - \dfrac{2}{3}\right) \div \dfrac{4}{5}$

21. $\left(\dfrac{4}{21}\right) \cdot \left(\dfrac{7}{8} + \dfrac{3}{4}\right) \div \dfrac{13}{14}$

22. $\dfrac{1}{6} + \left(\dfrac{1}{3} - \dfrac{3}{10}\right) \div \dfrac{7}{15}$

23. $\left(\dfrac{1}{2}\right)^3 - \left(\dfrac{2}{5}\right)\left(\dfrac{15}{22}\right) + \dfrac{1}{4}$

24. $\left(\dfrac{1}{3}\right)^2 + \left(\dfrac{5}{6} - \dfrac{3}{8}\right) \div \dfrac{5}{12}$

25. $\dfrac{8}{9} - \left(\dfrac{7}{10}\right)\left(\dfrac{2}{3}\right) \div \dfrac{4}{5}$

26. $\left(1\dfrac{1}{4} - \dfrac{2}{7} + \dfrac{1}{28}\right) \div \left(\dfrac{3}{13}\right)$

27. $\left(\dfrac{11}{12} - \dfrac{2}{3}\right) + \dfrac{15}{16} \div \left(\dfrac{1}{2}\right)$

1. _____

2. _____

3. _____

4. _____

5. _____

6. _____

7. _____

8. _____

9. _____

10. _____

11. _____

12. _____

13. _____

14. _____

15. _____

16. _____

17. _____

18. _____

19. _____

20. _____

21. _____

22. _____

23. _____

24. _____

25. _____

26. _____

27. _____

Name   Score

Write the fraction as a decimal.

1. $\dfrac{7}{100}$ 2. $\dfrac{353}{1000}$ 3. $\dfrac{9}{10}$

1. _____

2. _____

3. _____

Write the decimal as a fraction.

3. 0.501   4. 0.07   5. 0.8411

3. _____

4. _____

5. _____

Write the number in words.

7. 0.39   8. 0.81   9. 2.007

7. _____

_____

8. _____

_____

9. _____

_____

10. 26.379   11. 514.3118   12. 1078.00002

10. _____

_____

11. _____

_____

12. _____

_____

Write the number in standard form.

13. Eight hundred thirty-four thousandths   14. Fifty-two millionths

13. _____

14. _____

15. Six and one hundred one ten-thousandths   16. Eighty-seven and nine hundred six thousandths

15. _____

16. _____

17. Twenty-five and seven thousand two hundred ninety-three hundred-thousandths   18. Ninety-one and seventeen ten-thousandths

17. _____

18. _____

Name   Score

Place the correct symbol, < or >, between the two numbers.

1. 0.23   0.3   2. 0.45   0.5   1. _____

                                                                   2. _____

3. 4.54   4.45   4. 7.10   7.01   3. _____

                                                                   4. _____

5. 9.143   9.134   6. 0.091   0.101   5. _____

                                                                   6. _____

7. 0.4103 0.413   8. 0.25   0.256   7. _____

                                                                   8. _____

9. 0.63   0.063   10. 0.3   1.003   9. _____

                                                                   10. _____

11. 0.7   0.079   12. 0.86   0.859   11. _____

                                                                   12. _____

13. 3.025   3.25   14. 0.54   0.0054   13. _____

                                                                   14. _____

15. 2.907   2.097   16. 0.8555 0.86   15. _____

                                                                   16. _____

Write the given numbers in order from smallest to largest.

17. 0.0037, 0.037, 0.00037, 0.37   18. 0.851, 0.0086, 0.086, 0.86   17. _____

                                                                   18. _____

19. 0.49, 0.05, 0.5, 0.049   20. 0.11, 0.0001, 0.012, 0.21   19. _____

                                                                   20. _____

Name   Score

Round the number to the given place value.

1. 0.064 Tenths	2. 9.138 Tenths	1. _____
		2. _____
3. 26.349 Tenths	4. 96.4501 Tenths	3. _____
		4. _____
5. 65.34498 Hundredths	6. 13.01264 Hundredths	5. _____
		6. _____
7. 517.677 Hundredths	8. 792.246 Hundredths	7. _____
		8. _____
9. 2.09181 Thousandths	10. 6.27958 Thousandths	9. _____
		10. _____
11. 79.4625 Thousandths	12. 51.00439 Thousandths	11. _____
		12. _____
13. 0.04195 Ten-thousandths	14. 0.003642 Ten-thousandths	13. _____
		14. _____
15. 4.37628 Ten-thousandths	16. 16.111919 Ten-thousandths	15. _____
		16. _____
17. 0.249668 Hundred-thousandths	18. 0.009123 Hundred-thousandths	17. _____
		18. _____
19. 7.880102 Hundred-thousandths	20. 11.732405 Hundred-thousandths	19. _____
		20. _____
21. 1.49256 Nearest whole number	22. 3.60021 Nearest whole number	21. _____
		22. _____
23. 70.50648 Nearest whole number	24. 0.0045895 Millionths	23. _____
		24. _____

Name   Score

Solve.

1. A penny has a diameter of 19.05 mm. What is the diameter of the penny rounded to the nearest mm?

2. A penny has a thickness of 1.55 mm. What is the thickness of the penny to the nearest tenth of a mm?

1. _____

2. _____

3. A dime weighs 2.268 g. What is the weight of the dime rounded to the nearest hundredth of a gram?

4. The rolling pin was invented by Catherine Deiner in 1891. Alphabet blocks were invented by Adeline Whitney in 1882. Which invention was created first?

3. _____

4. _____

5. During their careers, Marshall Faulk scored 136 touchdowns and Emmitt Smith scored 145 touchdowns. Who has the greater number of touchdowns?

6. During their careers, John Elway passed for 51, 475 yards and Dan Marino passed for 61,361 yards. Who had the greater number of passing yards?

5. _____

6. _____

7. In his career, Joe Montana passed for 40,551 yards. What is the number of yards rounded to the nearest thousand yards?

8. One meter is equivalent to 3.2808 feet. How many feet are in one meter rounded to the nearest tenth of a foot?

7. _____

8. _____

9. One nautical mile is equivalent to 1.1508 miles. How many miles are in one nautical miles rounded to the nearest hundredth of a mile?

10. The average annual snowfall for Billings, Montana is 56.9 in. The average annual snowfall for Albany, New York is 64.4 in. In which city is the smaller amount of annual snowfall?

9. _____

10. _____

50

Name _____   Score _____

Add or subtract.

1. $4.825 + 31.7894 + 168.67$ 2. $25.25 + 7.4418 + 18.5$

3. $46.287 - 13.91$ 4. $23.031 - 17.61$

5. $6.841 + 54 + 59.3254$ 6. $85.0013 + 1.407 + 3.1114$

7. $145.03 - 8.2174$ 8. $650 - 56.413$

9. $6.421 + 52.118 + 3 + 0.0098$ 10. $4.46 + 2.3845 + 2.5 + 0.0231$

11. $14.1 - 11.7809$ 12. $43.001 - 19.875$

13. $0.0014 + 83.9 + 46 + 148.0908$ 14. $75.514 + 0.199 + 29 + 8.356$

15. $143.24 - 80.794$ 16. $9.08 - 6.324$

1. _____
2. _____
3. _____
4. _____
5. _____
6. _____
7. _____
8. _____
9. _____
10. _____
11. _____
12. _____
13. _____
14. _____
15. _____
16. _____

Evaluate the variable expression $x + y + z$ for the given values of x, y, and z.

17. $x = 2.7156$; $y = 45.08$;   18. $x = 5.52$; $y = 94.099$;   19. $x = 11$; $y = 77.29$;
     $z = 6.0406$ $z = 7.2148$ $z = 5.0531$

17. _____
18. _____
19. _____

Evaluate the variable expression $x - y$ for the given values of x and y.

20. $x = 0.452$; $y = 0.39$ 21. $x = 0.847$; $y = 0.25$ 22. $x = 9.406$; $y = 6.315$

20. _____
21. _____
22. _____

Name   Score

Solve.

1. An athlete bicycles 17.2 miles on Monday, 6.7 miles on Tuesday, and 12.4 miles on Wednesday. What was the total distance traveled for the three days?

2. The odometer of your car read 2456.2 miles. You drive 65.4 miles on Friday, 56.9 miles on Saturday, and 73.3 miles on Sunday. Find your odometer reading at the end of the three days.

1. _____

2. _____

3. You buy a toaster for $16.83. How much change do you receive from a $20.00 bill?

4. During an experiment, a chemist noted that the temperature of a solution rose 5.75° the first minute and then rose another 7.5° the second minute. If the temperature of the solution was 38.25° after the first two minutes, what was the original temperature of the solution?

3. _____

4. _____

5. You bought a 5.3-pound roast. After you trimmed the fat, 4.6 pounds of meat remained. How much fat did you trim from the roast?

6. A competitive swimmer beat the team's record time of 56.27 seconds in the 100-meter freestyle competition by 0.89 seconds. What is the new record time?

5. _____

6. _____

7. Use the formula $M = S - C$, where M is the markup on a consumer product, S is the selling price, and C is the cost of the product to the business, to find the markup on a product that cost a business $2,837.27 and has a selling price of $3,899.99.

8. Find the equity on a home that is valued at $327,000 when the homeowner has $186,249.20 in loans on the property. Use the formula $E = V - L$, where E is the equity, V is the value of the home, and L is the loan amount on the property.

7. _____

8. _____

Name   Score

Multiply.

1. $(0.5)(0.7)$
2. $(6.4)(0.3)$
3. What is 5.6 times 9?
4. Find the product of 0.28 and 0.6.
5. $(5.1)(4.5)$
6. $(0.96)(3.7)$
7. $(2.64)(0.03)$
8. $5.83(0.008)$
9. $(0.67)(0.41)$
10. $(52.9)(0.2)$
11. $(1.07)(0.066)$
12. $(24.8)(0.0019)$
13. $(7.673)(0.45)$
14. $(0.314)(0.061)$
15. 5.92×0.8
16. 0.76×0.6
17. 3.9×0.44
18. 8.21×10

1. ___
2. ___
3. ___
4. ___
5. ___
6. ___
7. ___
8. ___
9. ___
10. ___
11. ___
12. ___
13. ___
14. ___
15. ___
16. ___
17. ___
18. ___

Evaluate the expression for the given values of the variables.

19. $1000c$, when $c = 6.8235$
20. xy, when $x = 3.278$ and $y = 4.6$
21. ab, when $a = 2.073$ and $y = 9.5$
22. xy, when $x = 0.063$ and $y = 0.34$

19. ___
20. ___
21. ___
22. ___

Name _____   Score _____

Divide.

1. $5.64 \div 6$ 2. $2.24 \div 0.7$ 3. $40 \div 0.8$

4. $25.95 \div 0.3$ 5. $6.515 \div 5$ 6. $0.899 \div 2.9$

7. $1.288 \div 0.46$ 8. $42.3 \div 0.09$ 9. $0.1116 \div 0.012$

Divide. Round to the nearest tenth.

10. $73.85 \div 9.6$ 11. $0.473 \div 0.54$ 12. $1.265 \div 0.043$

Divide. Round to the nearest hundredth.

13. $4.724 \div 17$ 14. $8 \div 0.41$ 15. $36.597 \div 53.2$

Divide. Round to the nearest thousandth.

16. $0.0717 \div 0.9$ 17. $69.418 \div 83.5$ 18. $0.4728 \div 57.5$

Divide. Round to the nearest whole number.

19. $34.19 \div 36$ 20. $6.25 \div 0.4$ 21. $5.125 \div 0.073$

Evaluate the variable expression $x + y$ for the given values of x and y.

22. $x = 7.23; y = 10$ 23. $x = 2.898; y = 0.92$ 24. $x = 0.01878; y = 0.03$

1. _____
2. _____
3. _____
4. _____
5. _____
6. _____
7. _____
8. _____
9. _____
10. _____
11. _____
12. _____
13. _____
14. _____
15. _____
16. _____
17. _____
18. _____
19. _____
20. _____
21. _____
22. _____
23. _____
24. _____

Name _____ Score _____

Convert the fraction to a decimal. Place a bar over repeating digits of a repeating decimal. Round non-repeating decimals to the nearest thousandth.

1. $\dfrac{8}{9}$

2. $\dfrac{3}{14}$

3. $\dfrac{6}{11}$

4. $\dfrac{35}{8}$

5. $41\dfrac{3}{10}$

6. $1\dfrac{2}{23}$

7. $\dfrac{17}{18}$

8. $6\dfrac{5}{9}$

9. $10\dfrac{12}{21}$

1.	_____
2.	_____
3.	_____
4.	_____
5.	_____
6.	_____
7.	_____
8.	_____
9.	_____

Convert the decimal to a fraction.

10. 0.74

11. 0.375

12. 0.205

13. 4.138

14. 6.064

15. 3.35

16. $0.11\dfrac{1}{9}$

17. 4.81

18. 0.055

10.	_____
11.	_____
12.	_____
13.	_____
14.	_____
15.	_____
16.	_____
17.	_____
18.	_____

Place the correct symbol, < or >, between the two numbers.

19. 0.399 $\dfrac{2}{5}$

20. 0.433 $\dfrac{7}{16}$

21. $\dfrac{5}{9}$ 0.54

22. $\dfrac{1}{8}$ 0.124

23. $\dfrac{11}{15}$ 0.734

24. 0.589 $\dfrac{3}{5}$

19.	_____
20.	_____
21.	_____
22.	_____
23.	_____
24.	_____

Name _____   Score _____

Solve.

1. A sheet of plywood is 0.25 inch thick. Find the height of a stack of 150 sheets of plywood.

2. A shuttle bus transports students from a suburban college to work study jobs in the city and back again 5 times a day. If the distance between the college and the city is 11.3 miles, find the distance the shuttle bus travels in one day.

1. _____

2. _____

3. A car is bought for $3600 down and payments of $141.50 each month for 24 months. Find the total cost of the car.

4. You obtain a simple interest loan of $2450. The interest on the loan after one year is found by multiplying the loan amount by 0.12. Find the total amount due at the end of one year.

3. _____

4. _____

5. You pay $1284.72 per year in life insurance premiums. You pay the premiums in 12 equal monthly payments. Find the amount of each monthly payment.

6. Gasoline tax is $0.19 per gallon. Find the number of gallons used during a month in which $158.84 was paid in taxes.

5. _____

6. _____

7. A tax of $1.39 is paid on each hair dryer sold by a store. This month the total tax paid on hair dryers was $31.97. How many hair dryers were sold?

8. A jogger ran 6.8 miles in 42.16 minutes. What was the jogger's average time per mile?

7. _____

8. _____

Name _____ Score _____

Solve. Write the answer as a decimal.

1. $y + 2.67 = 5.19$

2. $x - 3.6 = 1.49$

1. _____

2. _____

3. $-6.7 = c - 13$

4. $-34 = x - 5.73$

3. _____

4. _____

5. $-4.1 = \dfrac{y}{5.3}$

6. $11 = z + 0.97$

5. _____

6. _____

7. $5.32r = -2.128$

8. $-97.6a = 204.96$

7. _____

8. _____

9. $\dfrac{x}{3} = -0.84$

10. $-4.02 = -\dfrac{z}{5}$

9. _____

10. _____

11. $\dfrac{t}{-3.1} = -6.7$

12. $-44 = -50x$

11. _____

12. _____

13. $0.813 = 3.998 + x$

14. $-7v = \dfrac{7}{8}$

13. _____

14. _____

15. $1.69 = -0.13t$

16. $-87.1x = 34.84$

15. _____

16. _____

Name _____   Score _____

Solve.

1. The average acceleration of an object is given by the formula $a = \dfrac{v}{t}$, where a is the average acceleration, v is the velocity and t is the time. Find the velocity after 7.3 s of an object whose acceleration is 16 ft/s^2.

2. The fundamental accounting equation is $A = L + S$, where A is the assets of a company, L is the liabilities of the company, and S is the stockholders' equity. Find the stockholders' equity in a company whose assets are $36.7 million and whose liabilities are $28.8 million.

1. _____

2. _____

3. The markup on an item in a store equals the difference between the selling price of the item and the cost of the item. Find the selling price of a package of golf balls for which the cost is $10.34 and the markup is $6.16.

4. The total of the monthly payments for a car lease is the product of the number of months of the lease and the monthly lease payment. The total of the monthly payments for a 48-month car lease is $17,976. Find the monthly lease payment.

3. _____

4. _____

5. The area of a rectangle is 152 in^2. If the width of the rectangle is 9.5 in., what is the length? Use the formula $A = LW$.

6. The length of a rectangle is 22 ft. If the area is 297 ft^2, what is the width of the rectangle? Use the formula $A = LW$.

5. _____

6. _____

Name   Score

Simplify.

1. $\sqrt{49}$ 2. $-\sqrt{16}$ 1. _____

                                                                       2. _____

3. $\sqrt{144}$ 4. $-\sqrt{121}$ 3. _____

                                                                       4. _____

5. $\sqrt{9+40}$ 6. $\sqrt{69+12}$ 5. _____

                                                                       6. _____

7. $\sqrt{64}+\sqrt{25}$ 8. $\sqrt{169}-\sqrt{64}$ 7. _____

                                                                       8. _____

9. $-4\sqrt{81}$ 10. $6\sqrt{25}-11$ 9. _____

                                                                       10. _____

11. $15\sqrt{4}-\sqrt{49}$ 12. $\sqrt{\dfrac{25}{64}}$ 11. _____

                                                                       12. _____

13. $\sqrt{\dfrac{4}{49}}$ 14. $\sqrt{\dfrac{1}{49}}-\sqrt{\dfrac{1}{81}}$ 13. _____

                                                                       14. _____

15. $\sqrt{\dfrac{1}{16}}+\sqrt{\dfrac{1}{25}}$ 16. $\sqrt{25}+\sqrt{1}$ 15. _____

                                                                       16. _____

Evaluate the expression for the given values of the variables.

17. $-5\sqrt{xy}$ 18. $9\sqrt{x+y}$ 19. $\sqrt{b^2-4ac}$ 17. _____
    for $x=2$ and $y=8$ for $x=54$ and $y=10$ for $a=2$, $b=7$
                                                                 and $c=-4$ 18. _____

                                                                                     19. _____

59

Name   Score

Approximate to the nearest ten-thousandth.

1. $\sqrt{17}$ 2. $7\sqrt{18}$ 3. $-9\sqrt{29}$ 1. _____

                                                                                                                    2. _____

                                                                                                                    3. _____

4. $-13\sqrt{42}$ 5. $4\sqrt{26}$ 6. $-2\sqrt{31}$ 4. _____

                                                                                                                    5. _____

                                                                                                                    6. _____

7. $5\sqrt{33}$ 8. $11\sqrt{15}$ 9. $-21\sqrt{6}$ 7. _____

                                                                                                                    8. _____

                                                                                                                    9. _____

Between what two whole numbers is the value of the radical expression?

10. $\sqrt{145}$ 11. $\sqrt{75}$ 12. $\sqrt{111}$ 10. _____

                                                                                                                    11. _____

                                                                                                                    12. _____

Simplify.

13. $\sqrt{24}$ 14. $\sqrt{60}$ 15. $\sqrt{125}$ 13. _____

                                                                                                                    14. _____

                                                                                                                    15. _____

16. $\sqrt{160}$ 17. $\sqrt{280}$ 18. $\sqrt{300}$ 16. _____

                                                                                                                    17. _____

                                                                                                                    18. _____

19. $\sqrt{252}$ 20. $\sqrt{50}$ 21. $\sqrt{126}$ 19. _____

                                                                                                                    20. _____

                                                                                                                    21. _____

Name _____   Score _____

Solve.

1. Use the formula $v = 3\sqrt{d}$, where v is the velocity in feet per second of a tsunami as it approaches land and d is the depth in feet of the water. Find the velocity of the tsunami when the depth of the water is 121 ft.

2. Use the formula $v = 3\sqrt{d}$, where v is the velocity in feet per second of a tsunami as it approaches land and d is the depth in feet of the water. Find the velocity of the tsunami when the depth of the water is 196 ft.

1. _____

2. _____

3. Use the formula $v = 3\sqrt{d}$, where v is the velocity in feet per second of a tsunami as it approaches land and d is the depth in feet of the water. Find the velocity of the tsunami when the depth of the water is 81 ft.

4. Use the formula $t = \sqrt{\dfrac{d}{16}}$, where t is the time in seconds that an object falls and d is the distance in feet that the object falls. If an object is dropped from a plane, how long will it take for the object to fall 121 ft?

3. _____

4. _____

5. Use the formula $t = \sqrt{\dfrac{d}{16}}$, where t is the time in seconds that an object falls and d is the distance in feet that the object falls. If an object is dropped from a plane, how long will it take for the object to fall 400 ft?

6. Use the formula $t = \sqrt{\dfrac{d}{16}}$, where t is the time in seconds that an object falls and d is the distance in feet that the object falls. If an object is dropped from a plane, how long will it take for the object to fall 256 ft?

5. _____

6. _____

Name

Score

Graph the number on the real number line.

1.   2

2.   6

3.   7

4.   1

Graph.

5.   the real numbers greater than 3

6.   the real numbers less than 3

7.   the real numbers greater than –1

8.   the real numbers greater than –4

9.   the real numbers less than 4

10.   the real numbers greater than –2

11.   the real numbers less than 0

12.   the real numbers greater than –5

15.   the real numbers between –1 and 2

16.   the real numbers between –4 and 4

Name   Score

For the given inequality, which numbers listed make the inequality true?

1. $x > -4$
   $-10.5, -8, -4, 3$

2. $x \leq 2$
   $-7.6, -3, 0, 4.2$

3. $x < 6$
   $-6, 1.9, 8, 9.6$

4. $x \geq 0$
   $-5, -0.3, 0, 6$

5. $x < -1$
   $-5, -1.1, -1, -0.9, 0.9$

6. $x > 3$
   $-2, 0, 3, 10$

1. _____

2. _____

3. _____

4. _____

5. _____

6. _____

Graph the solution set.

7. $x > 5$

8. $x \leq -3$

9. $x \geq 4$

10. $x < 3$

11. $x \geq -4$

12. $x \geq 0$

13. $x < -1$

14. $x > -2$

63

Name _____   Score _____

Solve.

1. Each sales representative for a
    company must sell at least 75,000
    units per year. Write an inequality
    for the number of units a sales
    representative must sell. Has a
    representative who has sold 69,000
    units this year met the sales goal?

2. A health official recommends a
    cholesterol level of less than 220
    units. Write an inequality for the
    acceptable cholesterol levels. Is a
    cholesterol level of 211 within the
    recommended levels?

1. _____

2. _____

3. A part-time student can take a
    maximum of 9 credit hours per
    semester. Write an inequality for
    the number of credit hours a part-
    time student can take. Does a
    student taking 10 credit hours
    fulfill the requirement for being
    a part-time student?

4. A service organization will receive
    a bonus of $250 for collecting
    more than 1,850 lb of aluminum
    cans during a collection drive.
    Write an inequality for the number
    of cans that must be collected in
    order to earn the bonus. If 1,856.5 lbs
    of aluminum cans are collected,
    will the organization receive the
    bonus?

3. _____

4. _____

5. Your monthly budget allows you
    to spend at most $2,250 per month.
    Write an inequality for the amount
    of money you can spend per month.
    Have you kept within your budget
    during a month in which you spent
    $2,199?

6. In order to get a B in a history
    course, you must earn more than
    80 points on the final exam. Write
    an inequality for the number of
    points you need to score on the
    final exam. Will a score of $79\frac{1}{2}$
    earn you a B in the course?

5. _____

6. _____

Name   Score

Simplify.

1. $2(5x)$ 2. $6(4x)$ 3. $-3(2a)$

4. $-3(-4y)$ 5. $-7(-8y)$ 6. $-6(2x^2)$

7. $-7(9x^2)$ 8. $\frac{1}{5}(5x^2)$ 9. $\frac{1}{4}(4x^2)$

10. $\frac{1}{9}(9a)$ 11. $\frac{1}{10}(10x)$ 12. $-\frac{1}{3}(-3x)$

13. $-\frac{1}{5}(-5a)$ 14. $-\frac{1}{10}(-10a)$ 15. $(7x)\left(\frac{1}{7}\right)$

16. $(15x)\left(\frac{1}{15}\right)$ 17. $(-8y)\left(-\frac{1}{8}\right)$ 18. $(-20n)\left(-\frac{1}{20}\right)$

19. $\frac{1}{5}(15x)$ 20. $\frac{1}{6}(24x)$ 21. $-\frac{1}{3}(12x)$

22. $-\frac{1}{6}(18x)$ 23. $-\frac{7}{8}(32a^2)$ 24. $-\frac{3}{5}(25x^2)$

25. $15y+(-15y)$ 26. $(-4y)+3+4y$ 27. $52+(-52)+m$

1. _____
2. _____
3. _____
4. _____
5. _____
6. _____
7. _____
8. _____
9. _____
10. _____
11. _____
12. _____
13. _____
14. _____
15. _____
16. _____
17. _____
18. _____
19. _____
20. _____
21. _____
22. _____
23. _____
24. _____
25. _____
26. _____
27. _____

Name _____   Score _____

Simplify.

1. $-(x+5)$

2. $3(2x-1)$

3. $4(2x-3)$

4. $-3(a+2)$

5. $-4(a+12)$

6. $-2(2y-5)$

7. $-4(3y-4)$

8. $3(-2x^2-12)$

9. $4(-5x^2-2)$

10. $-5(2y^2-3)$

11. $-7(3y^2-10)$

12. $-6(x^2+y^2)$

13. $-5(x^2-y^2)$

14. $-3(x^2-2y^2)$

15. $-2(5a^2-8b^2)$

16. $6(x^2+x-7)$

17. $3(x^2-2x-5)$

18. $-2(y^2+3y-5)$

19. $3(-a^2-a-4)$

20. $-7(-2x^2+4x+6)$

21. $-2(-3x^2+4x-1)$

22. $2(x^2+2xy-4y^2)$

23. $3(2x^2-3xy-4y^2)$

24. $-(2a^2+4a-3)$

25. $-(9b^2-5b+8)$

1. _____

2. _____

3. _____

4. _____

5. _____

6. _____

7. _____

8. _____

9. _____

10. _____

11. _____

12. _____

13. _____

14. _____

15. _____

16. _____

17. _____

18. _____

19. _____

20. _____

21. _____

22. _____

23. _____

24. _____

25. _____

Name _____   Score _____

Simplify.

1. $4x + 7x$ 2. $10x + 11x$ 3. $7a - 2a$ 1. _____

                                                                  2. _____

                                                                  3. _____

4. $-2a + 9a$ 5. $7ab - 5ab$ 6. $-10xy + 15xy$ 4. _____

                                                                  5. _____

                                                                  6. _____

7. $9xy - 11xy$ 8. $-13xy + 9xy$ 9. $-8ab + 13ab$ 7. _____

                                                                  8. _____

                                                                  9. _____

10. $-4mn + 4mn$ 11. $-\frac{1}{3}x - \frac{1}{4}x$ 12. $-\frac{3}{5}x + \frac{2}{10}x$ 10. _____

                                                                  11. _____

                                                                  12. _____

13. $\frac{1}{4}x^2 - \frac{7}{12}x^2$ 14. $\frac{5}{8}x - \frac{1}{2}x$ 15. $-\frac{2}{3}x^2 - \frac{3}{4}x^2$ 13. _____

                                                                  14. _____

                                                                  15. _____

16. $6a - 2a + 4a$ 17. $-3x^2 - 8x^2 - 2x^2$ 16. _____

                                                                  17. _____

18. $-2y^2 + 7y^2 - 6y^2$ 19. $6x - 5x + 2y$ 18. _____

                                                                  19. _____

20. $7y - 8x + 3x$ 21. $7y + 7x - 7y$ 20. _____

                                                                  21. _____

22. $2a - 5b - 7b + 4a$ 23. $9y - 11x - 6y + 4y$ 22. _____

                                                                  23. _____

24. $x^2 - 8x - 3x^2 + 7x$ 25. $-4b + 3a - 9b + 11a$ 24. _____

                                                                  25. _____

Name	Score

Simplify.

1. $5x - 3(2x + 7)$

2. $3a - (4a + 1)$

3. $9 - (10x - 4)$

4. $7 - 2(3x - 5)$

5. $6 - (8 + 5y)$

6. $4n - (6 - 3n)$

7. $3x - (10 - x)$

8. $2(x + 1) - 4(x - 6)$

9. $3(x - 3) - 2(x + 4)$

10. $8(y - 1) + 2(5 - 2y)$

11. $5(2y - 5) - 2(4 - y)$

12. $2(x + y) - 3(x - y)$

13. $3(a + b) - (a - 2b)$

14. $3[x - 3(x - 2)]$

15. $5[x + 3(x + 6)]$

16. $-3[2x + 3(2 - x)]$

17. $-4[x + 2(6 - x)]$

18. $-2[3x - (x + 8)]$

19. $-5[2x - (4x + 1)]$

20. $3x - 2[x - 3(5 - x)]$

21. $-6x + 4[x - 6(2 - x)]$

22. $-4x - 3[3x - 3(x + 6)] - 5$

23. $3a - 2[3b - (2b - a)] + 4b$

24. $3a - 3[b - (2b - a)] + 5b$

25. $3x + 2(x - 2y) + 4(2x - 5y)$

26. $4y - 2(y - 2x) + 3(6x - y)$

1. _____

2. _____

3. _____

4. _____

5. _____

6. _____

7. _____

8. _____

9. _____

10. _____

11. _____

12. _____

13. _____

14. _____

15. _____

16. _____

17. _____

18. _____

19. _____

20. _____

21. _____

22. _____

23. _____

24. _____

25. _____

26. _____

Name   Score

Add. Use a vertical format.

1. $(x^2 + 5x) + (-2x^2 - 3x)$

2. $(2y^2 - y) + (3y^2 + 4y)$

3. $(y^2 + 3y) + (-2y - 5)$

4. $(2x^2 + 5x) + (2x - 12)$

5. $(4x^2 + 7x) + (3x - 8)$

6. $(y^2 - 2y) + (-3y - 6)$

7. $(x^2 + 5x + 10) +$ $(2x^2 + 3x + 1)$

8. $(3x^2 + 4x + 9) +$ $(x^2 - x - 6)$

9. $(2y^3 - y^2 - 3) +$ $(-2y^3 - y^2 - 6)$

10. $(4y^3 + 2y^2 - 2) +$ $(-3y^3 - 2y - 1)$

11. $(3a^3 - 7a^2 - 3) +$ $(-3a^2 + 7a + 3)$

12. $(3y^3 + 2y^2 - y) +$ $(-y^3 - 5y - 2)$

Add. Use a horizontal format.

13. $(3x^2 + x) + (2x^2 + 5x)$

14. $(-2y^2 + 3y) + (3y^2 + 5y)$

15. $(5x^2 - 4xy) +$ $(2x^2 + 5xy - 2y^2)$

16. $(6x^2 - 3xy) +$ $(x^2 + 7xy - 3y^2)$

17. $(-5x^2 - 8x - 2) +$ $(3x^2 - x - 5)$

18. $(a^2 - 6a + 9) +$ $(3a^2 + 3a + 6)$

19. $(-7x^2 + 6x + 4) +$ $(2x^2 - x - 3)$

20. $(2a^2 - 9a + 5) +$ $(3a^2 + 3a + 6)$

21. $(-4x^2 + 3x + 1) +$ $(-x^2 - x - 2)$

22. $(4x^3 + 6x - 6) +$ $(9x^2 - 7x + 4)$

23. $(2y^2 - 3y + 9) +$ $(5y^3 - 5y)$

24. $(6x^3 + 6x - 3) +$ $(x^2 - 7x - 5)$

1. _____
2. _____
3. _____
4. _____
5. _____
6. _____
7. _____
8. _____
9. _____
10. _____
11. _____
12. _____
13. _____
14. _____
15. _____
16. _____
17. _____
18. _____
19. _____
20. _____
21. _____
22. _____
23. _____
24. _____

Name _____ Score _____

Subtract. Use a vertical format.

1. $(x^2 - 3x) - (x^2 - 7x)$
2. $(y^2 + 2y) - (y^2 + 8y)$
3. $(2y^2 - 3y) - (y^2 + 6)$

4. $(x^2 - 3x + 2) - (x^2 + 4x + 5)$
5. $(2x^2 + 3x - 1) - (x^2 + 5x + 8)$
6. $(x^2 + 3x + 4) - (2x^2 - 9x - 2)$

7. $(3x^3 + 6x + 3) - (-2x^2 + 3x + 2)$
8. $(-3x^2 - 2x + 3) - (2x^3 + x^2 + 3)$
9. $(3x^3 + 6x + 3) - (-2x^2 + 3x + 2)$

10. $(4x^2 + x - 2) - (5x^2 + 2x - 1)$
11. $(3y^3 + 2y^2 - y) - (-y^3 - 5y - 2)$
12. $(2x^3 - 5x + 6) - (x^3 - x + 7)$

Subtract. Use a horizontal format.

13. $(y^2 - 6xy) - (y^2 - 3xy)$
14. $(x^2 - 2xy) - (-2x^2 + 3xy)$
15. $(2y^2 - 8xy) - (y^2 - 6xy)$

16. $(6y^2 - y + 6) - (-3y^2 - 2y - 1)$
17. $(3x^2 + 2x - 2) - (x^2 + 5x - 6)$
18. $(2x^2 + 6x - 1) - (5x^2 - x - 2)$

19. $(-x^3 + x^2 - x) - (-x^2 - x - 7)$
20. $(3x^3 - 2x - 6) - (2x^2 + 6x - 1)$
21. $(2x^3 + 4x - 7) - (-x^2 + x - 13)$

22. $(3x^3 + 4x - 4) - (8x^2 - 6x - 1)$
23. $(-4x^2 + 3x - 1) - (-x^2 - x - 2)$
24. $(a^2 - 7a + 5) - (2a^3 - 6a - 1)$

1. _____
2. _____
3. _____
4. _____
5. _____
6. _____
7. _____
8. _____
9. _____
10. _____
11. _____
12. _____
13. _____
14. _____
15. _____
16. _____
17. _____
18. _____
19. _____
20. _____
21. _____
22. _____
23. _____
24. _____

Name   Score

Solve.

1. One side of a triangle is $3x$ meters. The other two sides have lengths of $(3x - 1)$ meters and $2x$ meters. Find the perimeter of the triangle. Use the formula $P = a + b + c$.

2. One side of a triangle is $(x - 2)$ inches. The other two sides have lengths of $(3x + 2)$ inches and $4x$ inches. Find the perimeter of the triangle. Use the formula $P = a + b + c$.

1. _____

2. _____

3. The distance from Modesto to Stockton is $(2x^2 - 4x + 2)$ kilometers. The distance from Stockton to Sacramento is $(3x^2 + 3x - 5)$ kilometers. Find the distance from Modesto to Sacramento.

4. The distance from New York to Chicago is $(6y^2 + y - 5)$ miles. The distance from Chicago to San Francisco is $(2y^2 + 8y + 2)$ miles. Find the distance from New York to San Francisco.

3. _____

4. _____

Use the formula $P = R - C$, where P is the profit, R is the revenue, and C is the cost.

5. A company's total monthly cost, in dollars, for manufacturing and selling n dishwashers per month is $65n + 850$. The company's revenue, in dollars, from selling all n dishwashers is $-0.1n^2 + 300n$. Express in terms of n the company's monthly profit.

6. A company's total monthly cost, in dollars, for manufacturing and selling n desks per month is $95n + 2000$. The company's revenue, in dollars, from selling all n desks is $-0.3n^2 + 400n$. Express in terms of n the company's monthly profit.

5. _____

6. _____

7. A company's total monthly cost, in dollars, for manufacturing and selling n sofas per month is $25n + 1500$. The company's revenue, in dollars, from selling all n sofas is $-0.7n^2 + 150n$. Express in terms of n the company's monthly profit.

8. A company's total monthly cost, in dollars, for manufacturing and selling n recliners per month is $50n + 600$. The company's revenue, in dollars, from selling all n recliners is $-0.5n^2 + 650n$. Express in terms of n the company's monthly profit.

7. _____

8. _____

Name _____   Score _____

Multiply.

1. $(x)(3x)$

2. $(-2y)(y)$

3. $(2x)(3x)$

4. $(4y^3)(4y^2)$

5. $(-2a^2)(-4a^3)$

6. $(4a^6)(-2a^4)$

7. $(xy^2)(x^2y^3)$

8. $(-3x^3)(4x^2y)$

9. $(-3a^3)(3a^3b^2)$

10. $(ab^3)(a^2b^2)$

11. $(4xy^2)(-2x^2y^3)$

12. $(-5ab^2)(-a^2b^3)$

13. $(xy^2z)(x^2y^3)$

14. $(ab^2)(a^2bc^3)$

15. $(xy^2z)(x^3y^2)$

16. $(-ab^2)(a^3b^5)$

17. $(x^2y^3)(-x^2y)$

18. $(-4y^3z)(-7y^4z^3)$

19. $(2x^2y)(-3x^2y)$

20. $(-a^2bc^2)(ab^2c)(-abc^3)$

21. $(x^2y^2z)(x^2y^3)(y^3z^2)$

22. $(-2ab)(3a^2b^2)(-4a^3b^3)$

23. $(2x^2y^4)(4xy)(-3x^2y^2)$

24. $(2a^2b^2)(-3ab^2)(a^4b^2)$

25. $(x^2y^3)(-2xy)(-4x^2y^4)$

26. $(3a^2b)(-2a^3b^2)(a^5b)$

27. $(2ab^2)(-3abc^2)(5a^2c)$

1. _____
2. _____
3. _____
4. _____
5. _____
6. _____
7. _____
8. _____
9. _____
10. _____
11. _____
12. _____
13. _____
14. _____
15. _____
16. _____
17. _____
18. _____
19. _____
20. _____
21. _____
22. _____
23. _____
24. _____
25. _____
26. _____
27. _____

Name   Score

Simplify.

1. $\left(x^2\right)^2$ 2. $\left(x^2\right)^3$ 3. $\left(y^2\right)^4$

4. $\left(x^5\right)^2$ 5. $\left(y^4\right)^3$ 6. $\left(-x^3\right)^2$

7. $(2x)^3$ 8. $(3y)^2$ 9. $(-2x)^3$

10. $\left(-3y^2\right)^2$ 11. $\left(xy^2\right)^2$ 12. $\left(x^2y^3\right)^4$

13. $\left(-2ab^2\right)^3$ 14. $\left(x^2y^2\right)^5$ 15. $\left(xy^3\right)^5$

16. $\left(x^4y^3\right)^4$ 17. $\left(2x^2y^3\right)^2$ 18. $\left(-a^2b^4\right)^3$

19. $\left(a^2b\right)^2\left(ab^2\right)^2$ 20. $(-2y)\left(3y^2\right)^3$ 21. $\left(a^2b\right)(ab)^2$

22. $(-3x)\left(-2x^2y^2\right)^2$ 23. $(-2y)\left(-3x^2y\right)^2$ 24. $\left(ab^2\right)\left(-2ab^3\right)^2$

25. $\left(-2a^2\right)\left(3ab^2\right)^3$ 26. $\left(2a^2b\right)^3\left(-3b^2\right)$ 27. $\left(-3ab^2\right)^3\left(-3ab\right)^2$

1. _____
2. _____
3. _____
4. _____
5. _____
6. _____
7. _____
8. _____
9. _____
10. _____
11. _____
12. _____
13. _____
14. _____
15. _____
16. _____
17. _____
18. _____
19. _____
20. _____
21. _____
22. _____
23. _____
24. _____
25. _____
26. _____
27. _____

73

Name   Score

Multiply.

1. $x(x+1)$ 2. $y(2-y)$ 3. $-x(x+2)$ 1. _____

                                                                         2. _____

                                                                         3. _____

4. $2a(a-1)$ 5. $3b(b+5)$ 6. $-2x^2(x-1)$ 4. _____

                                                                         5. _____

                                                                         6. _____

7. $-4y^2(y+6)$ 8. $-x^2(2x^2-3)$ 9. $2x(5x^2-2x)$ 7. _____

                                                                         8. _____

                                                                         9. _____

10. $3y(2y-y^2)$ 11. $(2x-3)x$ 12. $(2x-1)3x$ 10. _____

                                                                         11. _____

                                                                         12. _____

13. $-x^2y(x-y^2)$ 14. $-xy^2(2x-y)$ 15. $x(x^2-2x+1)$ 13. _____

                                                                         14. _____

                                                                         15. _____

16. $x(x^2+3x-2)$ 17. $-a(a^2-6a-1)$ 18. $-b(2b^2+3b-6)$ 16. _____

                                                                         17. _____

                                                                         18. _____

19. $x^2(2x^2-3x-2)$ 20. $y^3(-2y^2-3y+4)$ 21. $2y^2(-2y^2-5y+8)$ 19. _____

                                                                         20. _____

                                                                         21. _____

22. $3x^2(4x^2-2x+7)$ 23. $4x^2(5x^2-x-9)$ 24. $5y^2(-y^2+3y-6)$ 22. _____

                                                                         23. _____

                                                                         24. _____

25. $ab(a^2-3ab-4b^2)$ 26. $xy(x^2-2xy+2y^2)$ 27. $ab(a^2+5ab-7b^2)$ 25. _____

                                                                         26. _____

                                                                         27. _____

Name   Score

Multiply. Use the FOIL method.

1. $(y+2)(y+3)$ 2. $(a-2)(a+5)$ 3. $(b-5)(b+4)$ 1. _____

                                                                                                                 2. _____

                                                                                                                 3. _____

4. $(y+2)(y-7)$ 5. $(x+9)(x-4)$ 6. $(y-6)(y-2)$ 4. _____

                                                                                                                 5. _____

                                                                                                                 6. _____

7. $(a-2)(a-8)$ 8. $(x+11)(x-3)$ 9. $(2x+1)(x+6)$ 7. _____

                                                                                                                 8. _____

                                                                                                                 9. _____

10. $(y+1)(3y+2)$ 11. $(2x-3)(x+3)$ 12. $(5x-2)(x+3)$ 10. _____

                                                                                                                 11. _____

                                                                                                                 12. _____

13. $(2x-1)(3x-5)$ 14. $(2y+9)(y-1)$ 15. $(4y-7)(y+2)$ 13. _____

                                                                                                                 14. _____

                                                                                                                 15. _____

16. $(5a+2)(6a+1)$ 17. $(6a-13)(2a-5)$ 18. $(5a-9)(2a-7)$ 16. _____

                                                                                                                 17. _____

                                                                                                                 18. _____

19. $(3b+11)(5b-4)$ 20. $(3a+10)(4a-3)$ 21. $(x+y)(x+2y)$ 19. _____

                                                                                                                 20. _____

                                                                                                                 21. _____

22. $(2a+b)(a+2b)$ 23. $(2x-3y)(x-y)$ 24. $(a-3b)(2a+3b)$ 22. _____

                                                                                                                 23. _____

                                                                                                                 24. _____

25. $(4a-b)(2a+5b)$ 26. $(3x-5y)(3x+2y)$ 27. $(5x+2y)(6x+y)$ 25. _____

                                                                                                                 26. _____

                                                                                                                 27. _____

Name _____   Score _____

Simplify.

1. 6^{-2}

2. 2^0

3. 5^{-3}

4. a^{-5}

5. xy^{-2}

6. $x^3 x^{-3}$

7. $\dfrac{a^{-7}}{a^7}$

8. $\dfrac{x^{-1}y^2}{x}$

9. x^{-4}

10. $\left(a^{-1}b^2\right)^0$

11. $\dfrac{6x^2}{2x}$

12. $\dfrac{18x^5}{-6x^2}$

13. $\dfrac{a^7b^9}{a^9b^3}$

14. $\dfrac{x^8 y^6}{x^4 y^3}$

15. $\dfrac{3y}{(9y)^0}$

16. $\dfrac{1}{25a^{-2}}$

17. $\dfrac{1}{x^{-6}}$

18. $\dfrac{-4a^2}{2^2 a^2}$

19. $\dfrac{-60a^3b^7}{80ab^3}$

20. $\dfrac{15a^2b^5}{-90a^3b^3}$

21. $\dfrac{-6y^5}{33y^3z^3}$

22. $\dfrac{-16x^7 y^5}{-64x^3 y^2}$

23. $\dfrac{-b^3c^8}{bc^7}$

24. $\dfrac{x^2 y}{xy^4}$

1. _____
2. _____
3. _____
4. _____
5. _____
6. _____
7. _____
8. _____
9. _____
10. _____
11. _____
12. _____
13. _____
14. _____
15. _____
16. _____
17. _____
18. _____
19. _____
20. _____
21. _____
22. _____
23. _____
24. _____

Name _____   Score _____

Write the number in scientific notation.

1.   0.000000000675   2.   1,150,000,000   1. _____

                                                                  2. _____

3.   0.0000005635   4.   5,600,000   3. _____

                                                                  4. _____

5.   97,610,000,000   6.   40,600,000,000,000   5. _____

                                                                  6. _____

7.   59,600,000   8.   0.00001906   7. _____

                                                                  8. _____

Write the number in decimal notation.

9. 3.4×10^{-5} 10. 1.805×10^{14} 9. _____

                                                                  10. _____

11. 6.095×10^{7} 12. 3.4×10^{-8} 11. _____

                                                                  12. _____

13. 1.69×10^{-6} 14. 2.09×10^{-8} 13. _____

                                                                  14. _____

15. 4.56×10^{5} 16. 9.6×10^{11} 15. _____

                                                                  16. _____

Name   Score

Translate into a variable expression.

1.   the sum of 6 and x 2. y less 10   1. _____

                                                                                   2. _____

3. x less than 12   4. t increased by 6   3. _____

                                                                                   4. _____

5. a decreased by 1   6.   5 subtracted from y 5. _____

                                                                                   6. _____

7. z divided by 8   8. m multiplied by 11   7. _____

                                                                                   8. _____

9.   10 more than the square of x 10. 15 decreased by the cube of x 9. _____

                                                                                   10. _____

11. 5 times the sum of n and 8   12. the sum of two-thirds of x and 6 11. _____

                                                                                   12. _____

13. x increased by the product of 3 and x 14. the quotient of -6 and y 13. _____

                                                                                   14. _____

15. the product of -2 and t 16. the product of 2 and the sum of x and 8   15. _____

                                                                                   16. _____

17. the product of 5 and the total of y and 6 18. 10 divided by the sum of y and 5   17. _____

                                                                                   18. _____

19. 12 more than one-fourth of the   20. 16 more than the product of m and -1 19. _____
     square of y
                                                                                   20. _____

21. x increased by the quotient of x and 2 22. three-fifths of the product of w and 10   21. _____

                                                                                   22. _____

23. y decreased by the product of y and 2 24. the product of 7 and the total of x and 3   23. _____

                                                                                   24. _____

Name   Score

Translate into a variable expression. Then simplify.

1. a number subtracted from the product of four and the number

2. a number added to the product of five and the number

3. three less than the sum of a number and twelve

4. a number increased by the difference between eleven and the number

5. a number plus the sum of the number and seven

6. a number minus the sum of the number and five

7. the sum of two-fifths of a number and three-tenths of the number

8. the difference between three-eights of a number and one-fourth of the number

9. three less than the total of a number and nine

10. four more than the total of a number and three

11. four times the sum of five times a number and six

12. the difference between eleven times a number and seven times the number

13. twelve more than the sum of ten and a number

14. fifteen decreased by the sum of a number and five

15. three times the sum of two consecutive integers

16. nine more than the sum of two consecutive integers

17. one-half of the sum of two consecutive even integers

18. twenty more than a number added to the difference between the number and nine

19. a number subtracted from the product of the number and five

20. six times the sum of the square of a number and four

1. _____

2. _____

3. _____

4. _____

5. _____

6. _____

7. _____

8. _____

9. _____

10. _____

11. _____

12. _____

13. _____

14. _____

15. _____

16. _____

17. _____

18. _____

19. _____

20. _____

Name _____   Score _____

Solve.

1. Ten gallons of paint were poured into two containers of different sizes. Use one variable to express the amount poured into each container.

2. A wire 35 feet long was cut into two pieces of different lengths. Use one variable to express the lengths of the two pieces.

1. _____

2. _____

3. The speed of a commercial jet is three times the speed of a corporate jet. Express the speed of the commercial jet in terms of the speed of the corporate jet.

4. A banker divided $4200 between two accounts, one paying 4% annual interest and the second paying 3.5% annual interest. Express the amount invested in the 4% account in terms of the amount invested in the 3.5% account.

3. _____

4. _____

5. The base of a triangle is 4 feet longer than the height of the triangle. Express the base of the triangle in terms of the height of the triangle.

6. The length of a rectangular piece of poster board is 4 times the width. Express the length of the paper in terms of the width.

5. _____

6. _____

7. A board 8 feet long was cut into two pieces. Express the length of the longer piece in terms of the length of the shorter piece.

8. The length of a rectangular area rug is 3 feet less than twice the width. Express the length of the rug in terms of the width of the rug.

7. _____

8. _____

9. The number of cherry trees in an orchard is approximately one-fifth the number of apple trees in the orchard. Express the number of cherry trees in the orchard in terms of the number of apple trees in the orchard.

10. The number of adjunct faculty at a community college is 206 more than the number of full-time faculty. Express the number of adjunct faculty in terms of the number of full-time faculty at the community college.

9. _____

10. _____

Name   Score

Solve and check.

1. $x + 3 = 8$ 2. $x + 4 = 7$ 3. $a - 4 = 11$

4. $y - 5 = 9$ 5. $1 + a = 10$ 6. $m + 7 = 2$

7. $t + 8 = 0$ 8. $n - 3 = -1$ 9. $x - 7 = -4$

10. $x - 9 = -5$ 11. $x - 5 = -3$ 12. $x + 4 = 4$

13. $a - 2 = -7$ 14. $x - 4 = -2$ 15. $z + 7 = 1$

16. $t - 5 = -3$ 17. $9 + a = 15$ 18. $-6 = n + 2$

19. $3 = m - 7$ 20. $5 = -10 + b$ 21. $-9 = -3 + x$

22. $12 = -4 + a$ 23. $c + \dfrac{4}{5} = -\dfrac{1}{5}$ 24. $x - \dfrac{1}{8} = \dfrac{1}{8}$

25. $x - \dfrac{1}{5} = \dfrac{2}{5}$ 26. $b + \dfrac{1}{2} = -\dfrac{1}{3}$ 27. $x + \dfrac{1}{3} = -\dfrac{5}{6}$

1. _____
2. _____
3. _____
4. _____
5. _____
6. _____
7. _____
8. _____
9. _____
10. _____
11. _____
12. _____
13. _____
14. _____
15. _____
16. _____
17. _____
18. _____
19. _____
20. _____
21. _____
22. _____
23. _____
24. _____
25. _____
26. _____
27. _____

Name   Score

Solve and check.

1. $4x = 12$ 2. $5y = 30$ 3. $2a = -10$

4. $3a = -15$ 5. $-6m = 18$ 6. $-2x = -24$

7. $-5n = -35$ 8. $-48 = -8y$ 9. $-42 = 6a$

10. $-18 = -6y$ 11. $-\dfrac{y}{3} = 2$ 12. $-\dfrac{b}{4} = 1$

13. $\dfrac{2}{3}y = 8$ 14. $\dfrac{3}{5}x = 12$ 15. $-\dfrac{3}{4}d = 9$

16. $-\dfrac{4}{5}m = 12$ 17. $\dfrac{2x}{7} = 4$ 18. $-\dfrac{2z}{5} = 8$

19. $-\dfrac{7z}{8} = 14$ 20. $\dfrac{3n}{4} = 6$ 21. $\dfrac{5x}{6} = -5$

22. $\dfrac{-3z}{5} = 18$ 23. $-9 = -\dfrac{3x}{4}$ 24. $-12 = \dfrac{-3a}{4}$

25. $\dfrac{2x}{3} = 1$ 26. $6d - 3d = 6$ 27. $9x - 5x = 20$

1. _____
2. _____
3. _____
4. _____
5. _____
6. _____
7. _____
8. _____
9. _____
10. _____
11. _____
12. _____
13. _____
14. _____
15. _____
16. _____
17. _____
18. _____
19. _____
20. _____
21. _____
22. _____
23. _____
24. _____
25. _____
26. _____
27. _____

Name   Score

Solve and check.

1. $2x + 3 = 15$

2. $3y + 5 = 26$

3. $4a - 7 = 9$

4. $5m - 16 = 19$

5. $3 = 4a + 15$

6. $3x - 7 = -22$

7. $6n - 9 = -51$

8. $9 = 5 + 2d$

9. $17 = 9 + 4z$

10. $8 - c = 7$

11. $6 - 2w = -4$

12. $7 - 5x = -23$

13. $9 - 4t = 1$

14. $16 - 7x = 2$

15. $5y - 35 = 0$

16. $12 + 3b = 0$

17. $14 + 2m = 0$

18. $-3x + 7 = -5$

19. $-4d + 5 = -31$

20. $-9x - 4 = -22$

21. $-11x + 20 = -2$

22. $-15 = -12y + 21$

23. $3 = 7 - 4a$

24. $2 = 12 - 5n$

25. $-33 = -6b + 3$

26. $-7x + 2 = -26$

27. $-4x - 36 = 0$

1. _____

2. _____

3. _____

4. _____

5. _____

6. _____

7. _____

8. _____

9. _____

10. _____

11. _____

12. _____

13. _____

14. _____

15. _____

16. _____

17. _____

18. _____

19. _____

20. _____

21. _____

22. _____

23. _____

24. _____

25. _____

26. _____

27. _____

Name _____   Score _____

Solve.

Use the equation $P = 4s$, where P is the perimeter of a square, and s is the length of a side of the square.

1. Find the perimeter of a square when   2. Find the length of a side of a square   1. _____
    the length of a side is 15 cm.   when the perimeter is 96 cm.

                                                                                                 2. _____

Use the formula $L = P \cdot N$, where L is the loan amount, P is the monthly payment, and N is the number of months.

3. A loan of $1152 is to be repaid in   4. The monthly payment on a $6300 loan   3. _____
    12 equal monthly payments. Find the   is $150. Find the number of months
    monthly payment.   in which the loan is repaid.

                                                                                                 4. _____

Use the equation $V = \pi r^2 h$, where V is the volume of the cylinder, r is the radius of the cylinder, and h is the height of the cylinder. Use $\frac{22}{7}$ for π.

5. The radius of a cylinder if 14 cm, and   6. The volume of a cylinder is 20,790 cm^3,   5. _____
    20 cm is the height. Find the volume.   and the radius is 21 cm. Find the height
                                                   of the cylinder.

                                                                                                 6. _____

Use the equation $A = P + I$, where A is the value of the investment after one year, P is the original investment, and I is the increase in value of the investment.

7. The value of a stock investment after   8. The value of an investment in gold   7. _____
    one year was $4000. The original   after one year was $15,000. The increase
    investment was $2800. Find the increase   in value during the year was $3500. Find
    in value of the investment.   the amount of the original investment.

                                                                                                 8. _____

Use the equation $A = \frac{1}{2}bh$, where A is the area of a triangle, b is the base of the triangle, and h is the height of the triangle.

9. Find the area of a triangle when the   10. Find the height of a triangle when the   9. _____
    base is 24 cm and the height is 15 cm.   area is 360 cm^2 and the base is 36 cm.

                                                                                                 10. _____

Name   Score

Solve and check.

1. $7x + 4 = 3x + 32$	2. $5y + 1 = y + 17$	3. $10m + 2 = 9m + 10$
4. $4x - 3 = 2x + 7$	5. $8a - 9 = 2a + 15$	6. $11y - 3 = 8y - 9$
7. $12b - 3 = 4b - 27$	8. $14x - 1 = 3x - 23$	9. $6a - 4 = a - 19$
10. $4x + 3 = 13 - x$	11. $3x - 4 = -22 - 3x$	12. $5y - 1 = -17 - 3y$
13. $2b + 9 = 5b + 15$	14. $m + 3 = 3m + 11$	15. $5d - 3 = 7d + 9$
16. $5y - 7 = 2y - 7$	17. $4a + 11 = a + 11$	18. $5 - 4x = 7 - 2x$
19. $9 - 5n = 15 - 2n$	20. $4 + 7x = 12 + 3x$	21. $3x - 5 = 8x$
22. $2a - 15 = 7a$	23. $7m = 3m + 24$	24. $8y = 4y + 40$
25. $-2x - 3 = 3x + 7$	26. $-6a - 1 = 2a + 23$	27. $-7n + 4 = -4n - 14$

1. _____
2. _____
3. _____
4. _____
5. _____
6. _____
7. _____
8. _____
9. _____
10. _____
11. _____
12. _____
13. _____
14. _____
15. _____
16. _____
17. _____
18. _____
19. _____
20. _____
21. _____
22. _____
23. _____
24. _____
25. _____
26. _____
27. _____

Name   Score

Solve and check.

1. $3x + 2(x - 1) = 8$ 2. $5y + 3(y - 3) = 15$ 3. $8n - 5(2n + 1) = 7$ 1. _____

                                                                                              2. _____

                                                                                              3. _____

4. $10x - 3(2x + 5) = 13$ 5. $8m - 5(m - 2) = 7$ 6. $5a - (2a - 7) = 10$ 4. _____

                                                                                              5. _____

                                                                                              6. _____

7. $7n - 3(4n - 1) = 13$ 8. $3(2x - 1) - 2 = 13$ 9. $2(3b + 2) - 7 = 9$ 7. _____

                                                                                              8. _____

                                                                                              9. _____

10. $4(2 - 3y) + 3y = 26$ 11. $5(1 - 2x) + 5 = 80$ 12. $9x + 1 = 3(2x + 7) - 2$ 10. _____

                                                                                              11. _____

                                                                                              12. _____

13. $4y - 5 = 6 + 3(y - 1)$ 14. $11x + 4 = 2(3x - 2) - 7$ 15. $8 - 7x = 14 - (5x + 6)$ 13. _____

                                                                                              14. _____

                                                                                              15. _____

16. $2x - 3 = 5(3x + 2)$ 17. $4n - 9 = 3(2n + 5)$ 18. $4b + 2(3b - 5) = 2b + 6$ 16. _____

                                                                                              17. _____

                                                                                              18. _____

19. $x + 2(3x - 1) = 6x - 5$ 20. $3 - 2a = 10 - 3(2a - 3)$ 21. $8 - 4x = 11 - (5x + 6)$ 19. _____

                                                                                              20. _____

                                                                                              21. _____

22. $4b + 3(2b - 1) = 2b + 13$ 23. $x + 2(3x - 2) = 6x - 2$ 24. $2y - 5 = 4(y - 7) + 5$ 22. _____

                                                                                              23. _____

                                                                                              24. _____

25. $3a - 4 = 3(2a + 1) + 5$ 26. $4 - (8 - 5x) = 3x - 8$ 27. $6 - (4 - 7x) = 3x + 2$ 25. _____

                                                                                              26. _____

                                                                                              27. _____

Name   Score

Solve.

Use the level system equation $F_1 \cdot x = F_2 \cdot (d - x)$.

1. A lever is 24 ft long. A force of 36 lb is applied to one end of the lever and a force of 12 lb is applied to the other end. Find the location of the fulcrum when the system balances.

2. A lever is 16 ft long. A force of 50 lb is applied to one end of the lever and a force of 30 lb is applied to the other. Find the location of the fulcrum when the system balances.

1. _____

2. _____

3. A lever is 12 ft long. At a distance of 8 ft from the fulcrum, a force of 24 lb is applied. How large a force must be applied to the other end of the lever so that the system balances?

4. A lever is 20 ft long. At a distance of 5 ft from the fulcrum, a force of 24 lb is applied. How large a force must be applied to the other end of the lever so that the system will balance?

3. _____

4. _____

Use the equation $S = C + R \cdot C$, where S is the selling price, C is the cost, and R is the mark-up rate.

5. A clothing store uses a mark-up rate of 40%. Find the selling price of a jacket which costs $25.

6. A store manager uses a mark-up rate of 25% on all frozen foods. Find the selling price of a frozen lobster which costs $12.

5. _____

6. _____

7. A jewelry store uses a mark-up rate of 35%. Find the cost of a ring which sells for $270.

8. A gift shop uses a mark-up rate of 50% on all lamps. Find the cost of a lamp which sells for $105.

7. _____

8. _____

Use the equation $m_1 \cdot (T_1 - T) = m_2 \cdot (T - T_2)$ to find the final temperature of the water. m_1 is the quantity of the water at the hotter temperature, T_1 is the temperature of the hotter water, m_2 is the quantity of water at the cooler temperature, T_2 is the temperature of the cooler water, and T is the final temperature of the water after mixing.

9. A chemist mixes 600 g of water at 60°C with 200g of water at 25°C. Find the final temperature of the water after mixing.

10. A chemist mixes 400g of water at 70°C with 80g of water at 15°C. Find the final temperature of the water after mixing.

9. _____

10. _____

Name _____   Score _____

Translate into an equation and solve.

1. The sum of a number and nine is fifteen. Find the number.

2. Two more than a number is eleven. Find the number.

1. _____
2. _____

3. The difference between a number and six is ten. Find the number.

4. One-half of a number is nine. Find the number.

3. _____
4. _____

5. Five more than four times a number is thirteen. Find the number.

6. Two times the sum of a number and three is twelve. Find the number.

5. _____
6. _____

7. Eight less than three times a number is equal to seven. Find the number.

8. The quotient of a number and six is the opposite of two. Find the number.

7. _____
8. _____

9. The sum of seven times a number and three is the opposite of eighteen. Find the number.

10. The sum of five times a number and the number is twenty-four. Find the number.

9. _____
10. _____

11. The difference between six times a number and three times the number is twelve. Find the number.

12. Thirty is five minus the product of five and a number. Find the number.

11. _____
12. _____

13. Nine is one more than the product of two and a number. Find the number.

14. The sum of two numbers is twenty-one. Twice the larger is four times the smaller number. Find the two numbers.

13. _____
14. _____

15. Five times a number is equal to eighteen more than twice the number. Find the number.

16. Four times the sum of a number and three equals two less than five times the number. Find the number.

15. _____
16. _____

Name _____   Score _____

Write an equation and solve.

1. The retail selling price of a smoke detector is $36. The price is $22 more than the cost of the smoke detector. Find the cost of the smoke detector.

2. The sale price of a pair of boots is $64. This is $16 less than the original price. Find the original price.

1. _____

2. _____

3. Due to depreciation, the value of a motorcycle is now $2400. This is three-fourths of its original value. Find the original value.

4. An engineer's salary is $54,000. This is twice the engineer's salary twelve years ago. Find the engineer's salary twelve years ago.

3. _____

4. _____

5. An appliance center sold four times as many color TV sets as it sold black and white TV sets. The number of TV sets sold last year was 1200. How many color TV sets were sold?

6. The sum of the angles of a triangle is 180°. The measure of the first angle is 60°. The measure of the second angle is twice the measure of the third angle. Find the measures of the second and third angles.

5. _____

6. _____

7. An air conditioning repair bill was $131. This included $75 for parts and $28 for each hour of labor. Find the number of hours of labor.

8. A driver drove for 12 hours. During the first 8 hours the driver averaged 50 mph. If the driver drove 560 miles, what was the average speed during the last 4 hours of the trip?

7. _____

8. _____

9. The perimeter of a triangle is 180 cm. The length of the first side is twice the length of the second side. The length of the third side is three times the length of the second side. Find the length of each side.

10. The engine speed of a car is 2400 rpm (revolutions per minute). This is equal to 500 rpm less than twice the drive shaft speed. Find the drive shaft speed.

9. _____

10. _____

Name   Score

Graph.

1. Graph the ordered pairs $(-1, 2)$, $(2, -4)$, $(-1, 3)$, and $(0, 1)$.

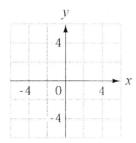

2. Graph the ordered pairs $(4, -2)$, $(-2, -2)$, $(-2, 0)$, and $(2, -1)$.

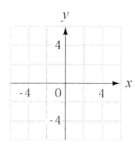

3. Graph the ordered pairs $(1, 1)$, $(1, -4)$, $(-2, 1)$, and $(1, -2)$.

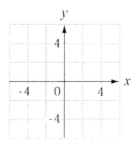

4. Graph the ordered pairs $(-3, 4)$, $(-2, 2)$, $(2, -3)$, and $(4, 0)$.

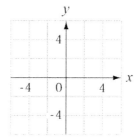

5. Graph the ordered pairs $(2, -3)$, $(-1, -2)$, $(0, 1)$, and $(4, 0)$.

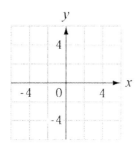

6. Graph the ordered pairs $(4, 1)$, $(-3, -2)$, $(2, 2)$, and $(0, -1)$.

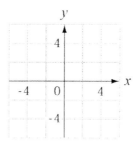

7. Find the coordinates of each of the points.

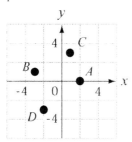

8. Find the coordinates of each of the points.

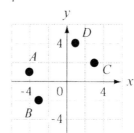

7. _____

8. _____

Name   Score

Solve.

1.   The table shows the number of U.S. households with a net worth of $1 million or more every
     three years from 1989 to 2001. Also shown is the millionaire households as a percent of all
     households in the United States. Graph the scatter diagram for these data.

Millionaire households (in millions), x	3.0	3.1	3.4	4.1	5.0
Percent of all households, y	3.2	3.2	3.4	4.0	4.7

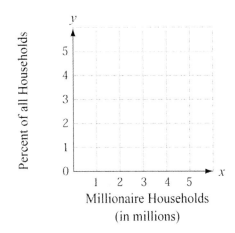

2.   As shown in the table below, the percent of households with telephones has increased over the
     years. Draw a scatter diagram for these data.

Year, x	'20	'30	'40	'50	'60	'70	'80	'90	'00
Percent, y	35	41	37	62	78	90	93	95	97

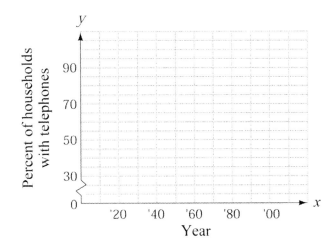

Name   Score

1. Is (2, 3) a solution of
    $y = -x + 5$?

2. Is (1, –2) a solution of
    $y = x + 1$?

3. Is (–2, 4) a solution of
    $y = \frac{1}{2}x + 3$?

4. Is (2, –2) a solution of
    $y = -2x + 2$?

5. Is (3, 2) a solution of
    $y = x - 1$?

6. Is (–4, 2) a solution of
    $y = -\frac{3}{2}x + 2$?

7. Is (0, 3) a solution of
    $y = \frac{3}{5}x + 3$?

8. Is (–1, 0) a solution of $y = -x$?

9. Is (0, 0) a solution of $y = 2x - 3$?

10. Is (–1, –1) a solution of $y = x$?

11. Find the ordered pair solution of
    $y = 2x - 1$ corresponding to $x = 2$.

12. Find the ordered pair solution of
    $y = 3x + 2$ corresponding to $x = -3$.

13. Find the ordered pair solution of
    $y = \frac{1}{2}x - 2$ corresponding to $x = 4$.

14. Find the ordered pair solution of
    $y = \frac{2}{5}x - 3$ corresponding to $x = 10$.

15. Find the ordered pair solution of
    $y = -2x + 1$ corresponding to $x = -3$.

16. Find the ordered pair solution of
    $y = \frac{3}{5}x - 6$ corresponding to $x = 5$.

17. Find the ordered pair solution of
    $y = \frac{3}{4}x + 6$ corresponding to $x = -4$.

18. Find the ordered pair solution of
    $y = -\frac{1}{3}x - 1$ corresponding to $x = 6$.

1. _____
2. _____
3. _____
4. _____
5. _____
6. _____
7. _____
8. _____
9. _____
10. _____
11. _____
12. _____
13. _____
14. _____
15. _____
16. _____
17. _____
18. _____

Name   Score

Graph.

1. $y = 2x - 5$ 2. $y = -2x + 4$

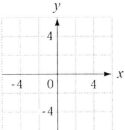

3. $y = \dfrac{1}{4}x$ 4. $y = -4x$

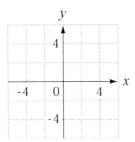

5. $y = \dfrac{1}{3}x + 1$ 6. $y = \dfrac{2}{3}x - 1$

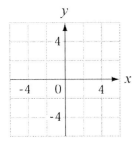

7. $y = 2x - 6$ 8. $y = -x + 1$

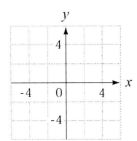

Name _____   Score _____

Convert.

1.   196 cm = _____ m   **2.**   243 mm = _____ cm   **3.**   4.218 km = _____ m   **1.** _____

                                                                                                                                 2. _____

                                                                                                                                 3. _____

4.   9.63 m = _____ cm   **5.**   0.345 g = _____ mg   **6.**   0.72 kg = _____ g   **4.** _____

                                                                                                                                 5. _____

                                                                                                                                 6. _____

7.   254 mg = _____ g   **8.**   2754 g = _____ kg   **9.**   2347 L = _____ kl   **7.** _____

                                                                                                                                 8. _____

                                                                                                                                 9. _____

10.   2700 ml = _____ L   **11.**   3.31 kl = _____ L   **12.**   0.0032 L = _____ ml   **10.** _____

                                                                                                                                 11. _____

                                                                                                                                 12. _____

Solve.

13.   A brick weighs 450 g. Find the   **14.**   How many VCRs weighing 12 kg   **13.** _____
           weight in kilograms of a load of   each can be placed on a shelf with
           600 bricks.   a maximum load limit of 100 kg?

                                                                                                                                 14. _____

15.   It is recommended that 1 kg of   **16.**   A concession stand at the fair sold   **15.** _____
           fertilizer be used for every 15 trees   1038 medium soft drinks. Each medium
           in an orange grove. Find the amount   soft drink cup contains 540 ml. How
           of fertilizer required for 1800 trees.   many liters of soft drink were sold?

                                                                                                                                 16. _____

17.   A motorist bought 40 liters of gasoline.   **18.**   A food processor used 125 ml of   **17.** _____
           The cost of the gasoline was $15.40.   vinegar in each jar of pickles. How
           Find the cost of one liter of gasoline.   many liters of vinegar are used when
                                                                          600 jars of pickles are prepared?

                                                                                                                                 18. _____

Name   Score

Write the comparison as a ratio in simplest form using a fraction, a colon, and the word *to*.

1. 3 yards to 7 yards   2.   6 minutes to 16 minutes   1. _____

                                                                                   2. _____

3. 3 cents to 21 cents   4.   9 hours to 6 hours   3. _____

                                                                                   4. _____

5. 9 pounds to 10 pounds   6.   60 cents to 24 cents   5. _____

                                                                                   6. _____

Solve.

7. In Rhode Island, the average summer and   8. For one month, a family spent $320 for   7. _____
    winter temperatures are 72°F and 28°F,   food, and $720 for rent. Find the ratio of
    respectively. Find the ratio of the summer   the amount spent for food to the amount
    temperature to the winter temperature.   spent for rent.   8. _____

Write as a rate in simplest form.

9. 182 miles on 6 gallons of gasoline   10. $310 for 15 toasters   9. _____

                                                                                   10. _____

11. 4 tablets in 24 hours   12. 675 clams for 225 people   11. _____

                                                                                   12. _____

13. 272 place settings on 34 tables   14. 868 seats in 8 lecture halls   13. _____

                                                                                   14. _____

Write as a unit rate.

15. 126 gallons in 15 minutes   16. 204 heartbeats in 3 minutes   15. _____

                                                                                   16. _____

17. 352 words in 5.5 minutes   18. $5000 for 4 partners   17. _____

                                                                                   18. _____

Solve.

19. During a city-wide power outage, the   20. A secretary typed a letter with 342   19. _____
    police station logged 651 calls in 5.25   words in 6 minutes. Find the number of
    hours. How many calls did the police   words typed in 1 minute.
    station receive per hour?   20. _____

95

Name _____   Score _____

Convert.

1. $9\frac{1}{3}$ ft. = _____ in.

2. 60 in. = _____ ft

3. $4\frac{2}{3}$ yd = _____ ft

4. 30 ft = _____ yd

5. $5\frac{1}{2}$ yd = _____ in.

6. 96 in. = _____ yd

7. 3 mi = _____ ft

8. $2\frac{1}{4}$ yd = _____ in.

9. 32 oz = _____ lb

10. 5 tons = _____ lb

11. 8000 lb = _____ tons

12. $\frac{3}{5}$ tons = _____ lb

13. 14 lb = _____ oz

14. 6400 lb = _____ tons

15. 90 oz = _____ lb

16. $9\frac{3}{4}$ lb = _____ oz

17. 24 fl oz = _____ c

18. 7 c = _____ pt

19. $2\frac{1}{2}$ pt = _____ c

20. 10 pt = _____ qt

21. 12 qt = _____ gal

22. 6 gal = _____ qt

23. $6\frac{1}{2}$ qt = _____ pt

24. 4 qt = _____ c

1. _____
2. _____
3. _____
4. _____
5. _____
6. _____
7. _____
8. _____
9. _____
10. _____
11. _____
12. _____
13. _____
14. _____
15. _____
16. _____
17. _____
18. _____
19. _____
20. _____
21. _____
22. _____
23. _____
24. _____

Name _____   Score _____

Solve.

1. A piece of plastic pipe $4\frac{1}{2}$ feet long is cut into four equal pieces. How long is each piece?

2. Forty-two yards of crepe paper were used for decorating. How many feet of crepe paper were used?

1. _____

2. _____

3. Thirty concrete blocks each 9 in. long, are laid end to end to make the foundation for a wall. Find the length of the wall in feet.

4. A brick weighs $2\frac{1}{2}$ lb. Find the weight of a load of 300 bricks.

3. _____

4. _____

5. A 9-by-9-inch tile weighs 8 oz. Find the weight in pounds of a package of 96 tiles.

6. Find the weight in pounds of 48 bars of soap. Each bar weighs 9 ounces.

5. _____

6. _____

7. A breakfast bar sold 88 cartons of orange juice in one day. Each carton contained 1 cup of juice. How many quarts of orange juice were sold that day?

8. One hundred twelve people attended the opening of the art exhibit. Assume that each person drank a cup of punch. How many gallons of punch were served?

7. _____

8. _____

9. A recipe calls for 10 ounces of tomato sauce. How many cups of tomato sauce will be needed if the recipe is tripled?

10. A can of pineapple juice contains 28 oz. Find the number of pints of pineapple juice in a case of 12 cans.

9. _____

10. _____

Name _____   Score _____

Convert. Round to the nearest hundredth.

1. Convert a 200-yard race to meters.

2. Find the weight in kilograms of a 187-pound person.

3. Find the number of liters in 6 gal of coffee.

4. Find the number of milliliters in 3 cups of milk.

5. Express 60 mi/h in kilometers per hour.

6. Grapes cost $1.49 per pound. Find the cost per kilogram.

7. Redwood stain costs $12.00 per gallon. Find the cost per liter.

8. The distance from Los Angeles to Atlanta is 2200 mi. Convert this distance to kilometers.

9. Find the number of gallons in 8 L of antifreeze.

10. Find the distance of a 2500-meter race in feet.

11. Find the weight in ounces of 500 g of cereal.

12. How many gallons does a 64-liter tank hold?

13. Find the width of a 40 mm piece of tape in inches.

14. A bottle of ketchup contains 906 ml. Find the number of pints in 906 ml.

15. Express 76 km/h in miles per hour.

16. A ham costs $4.30 per kilogram. Find the cost per pound.

1. _____

2. _____

3. _____

4. _____

5. _____

6. _____

7. _____

8. _____

9. _____

10. _____

11. _____

12. _____

13. _____

14. _____

15. _____

16. _____

Name   Score

Determine if the proportion is true or not true.

1. $\dfrac{6}{7} = \dfrac{12}{14}$

2. $\dfrac{15}{9} = \dfrac{25}{15}$

3. $\dfrac{9}{18} = \dfrac{12}{24}$

4. $\dfrac{606 \text{ words}}{10 \text{ minutes}} = \dfrac{302 \text{ words}}{5 \text{ minutes}}$

5. $\dfrac{48 \text{ cents}}{4 \text{ hours}} = \dfrac{60 \text{ cents}}{5 \text{ hours}}$

6. $\dfrac{282 \text{ houses}}{47 \text{ acres}} = \dfrac{188 \text{ houses}}{33 \text{ acres}}$

7. $\dfrac{450 \text{ gallons}}{60 \text{ minutes}} = \dfrac{180 \text{ gallons}}{24 \text{ minutes}}$

8. $\dfrac{6200 \text{ words}}{40 \text{ pages}} = \dfrac{7750 \text{ words}}{50 \text{ pages}}$

9. $\dfrac{375 \text{ cars}}{250 \text{ people}} = \dfrac{1218 \text{ cars}}{312 \text{ people}}$

1. _____

2. _____

3. _____

4. _____

5. _____

6. _____

7. _____

8. _____

9. _____

Solve. Round to the nearest hundredth.

10. $\dfrac{3}{5} = \dfrac{n}{10}$

11. $\dfrac{n}{8} = \dfrac{20}{32}$

12. $\dfrac{14}{24} = \dfrac{7}{n}$

13. $\dfrac{8}{n} = \dfrac{9}{27}$

14. $\dfrac{13}{36} = \dfrac{39}{n}$

15. $\dfrac{n}{18} = \dfrac{5}{9}$

16. $\dfrac{42}{15} = \dfrac{n}{12}$

17. $\dfrac{n}{8} = \dfrac{9}{12}$

18. $\dfrac{18}{n} = \dfrac{96}{132}$

19. $\dfrac{35}{n} = \dfrac{22}{11}$

20. $\dfrac{90}{n} = \dfrac{36}{3}$

21. $\dfrac{8}{18} = \dfrac{2}{n}$

22. $\dfrac{n}{15} = \dfrac{0.8}{5.6}$

23. $\dfrac{1.8}{18} = \dfrac{n}{12}$

24. $\dfrac{3.4}{20} = \dfrac{5.1}{n}$

10. _____

11. _____

12. _____

13. _____

14. _____

15. _____

16. _____

17. _____

18. _____

19. _____

20. _____

21. _____

22. _____

23. _____

24. _____

Name _____   Score _____

Solve. Round to the nearest hundredth.

1. A life insurance policy costs $4.52 for every $1000 of insurance. At this rate, what is the cost for $20,000 worth of life insurance?

2. A liquid plant food is prepared by using one gallon of water for each 1.5 teaspoon of plant food. At this rate, how many teaspoons of plant food are required for 7 gallons of water?

1. _____

2. _____

3. A $19.75 sales tax is charged for a $395 purchase. At this rate, what is the sales tax for a $621 purchase?

4. The scale on the plans for a new office building is 1 inch equals 4 feet. How long is a room that measures $8\frac{1}{2}$ inches on the drawing?

3. _____

4. _____

5. A stock investment of 150 shares paid a dividend of $555. At this rate, what dividend would be paid on 280 shares of stock?

6. A bank demands a loan payment of $18.95 each month for every $1000 borrowed. At this rate, what is the monthly payment for a $6000 loan?

5. _____

6. _____

7. A transistor company expects that 3 out of 245 transistors will be defective. How many defective transistors will be found in a batch of 184,485 transistors?

8. A department store makes a profit of $15.20 on every 25 rolls of film sold. How much profit is made if 16 rolls of film are sold?

7. _____

8. _____

9. For every 10 people who work in a city, 7 of them commute by public transportation. If 34,600 people work in the city, how many of them do not take public transportation?

10. For every 15 gallons of water pumped into the holding tank, 8 gallons were pumped out. After 930 gallons had been pumped in, how much water remained in the tank?

9. _____

10. _____

Name   Score

Solve.

1. Find the constant of variation when y varies directly as x and $y = 18$ when $x = 3$.

2. Find the constant of variation when q varies directly as the square of p and $q = 28$ when $p = 2$.

1. _____

2. _____

3. Given A varies directly as B and $A = 30$ when $B = 6$, find A when $B = 7$.

4. Given that D is directly proportional to the square of C and $D = 40$ when $C = 2$, find D when $C = 3$.

3. _____

4. _____

5. Given V varies directly as U, and $V = 48$ when $U = 8$, find V when $U = 7$.

6. Find the constant of variation when f varies directly as the square of h, and $f = 72$ when $h = 3$.

5. _____

6. _____

7. A worker's wage (w) is directly proportional to the number of hours (h) worked. If $80.50 is earned for 7 h, how much is earned for working 28 h?

8. The distance (d) a spring will stretch varies directly as the force (F) applied to the spring. If a force of 15 lb is required to stretch a spring 5 in., what force is required to stretch the spring 7 in.?

7. _____

8. _____

9. The number of words typed (w) is directly proportional to the time (t) spent typing. A typist can type 216 words in 3 min. Find the number of words typed in 16 min.

10. The stopping distance (s) of a car varies directly as the square of its speed (v). If a car traveling at 45 mph requires 137.7 ft to stop, find the stopping distance for a car traveling at 70 mph.

9. _____

10. _____

11. The distance (d) an object falls is directly proportional to the square of the time (t) of the fall. If an object falls a distance of 128 ft 2 s, how far will the object fall in 8 s?

12. The distance traveled (d) varies directly as the time (t) of travel, assuming that the speed is constant. If it takes 40 min to travel 30 mi, how many hours would it take to travel 234 mi?

11. _____

12. _____

Name   Score

Solve.

1. Find the constant of variation when y varies inversely as x and $y = 10$, when $x = 6$.

2. Find the constant of proportionality when t varies inversely as s, and $t = 0.4$ when $s = 15$.

3. Find the constant of variation when w varies inversely as the square of v, and $w = 7$ when $v = 4$.

4. If y varies inversely as x, and $y = 35$ when $x = 3$, find y when $x = 15$.

5. If L varies inversely as W and $L = 18$ when $W = 6$, find L when $W = 4$.

6. If y varies inversely as the square of x, and $y = 50$ when $x = 3$, find y when $x = 5$.

7. The length (L) of a rectangle of fixed area varies inversely as the width (W). If the length of the rectangle is 9 ft when the width is 6 ft, find the length of the rectangle when the width is 2 ft.

8. The time (t) of travel of an automobile trip varies inversely as the speed (v). At an average speed of 70 mph, a trip took 3 h. The return trip took 3.5 h. Find the average speed of the return trip.

9. A company that produces personal computers has determined that the number of computers it cal sell (S) is inversely proportional to the price (P) of the computer. Seventeen hundred computers can be sold if the price is $1,700. How many computers can be sold if the price is $2,000?

10. The speed (s) of a gear varies inversely as the number of teeth (t). If a gear that has 50 teeth makes 16 rpm (revolutions per minute), how many revolutions per minute will a gear that has 40 teeth make?

11. The intensity (I) of a light source is inversely proportional to the square of the distance (d) from the source. If the intensity is 25 lumens at a distance of 6 ft, what is the intensity at a distance of 4 ft?

12. The repulsive force (f) between the north poles of two magnets is inversely proportional to the square of the distance (d) between them. If the repulsive force is 20 lbs when the distance is 6 in., find the repulsive force when the distance is 2.5 in.

1. _____

2. _____

3. _____

4. _____

5. _____

6. _____

7. _____

8. _____

9. _____

10. _____

11. _____

12. _____

Name _____   Score _____

Write as a fraction and as a decimal.

1. 39%

2. 64%

3. 125%

4. 26%

5. 85%

6. 20%

7. 450%

8. 19%

9. 55%

Write as a fraction.

10. $7\frac{8}{9}\%$

11. $7\frac{2}{3}\%$

12. $25\frac{4}{5}\%$

13. $64\frac{1}{2}\%$

14. $43\frac{1}{3}\%$

15. $99\frac{3}{4}\%$

Write as a decimal.

16. 67.5%

17. 34.07%

18. 57.9%

19. 40%

20. 13.89%

21. 2.01%

1. _____

2. _____

3. _____

4. _____

5. _____

6. _____

7. _____

8. _____

9. _____

10. _____

11. _____

12. _____

13. _____

14. _____

15. _____

16. _____

17. _____

18. _____

19. _____

20. _____

21. _____

Name _____ Score _____

Write as a percent.

1. 0.32   2. 0.96   3. 0.04

4. 1.97   5. 2.14   6. 0.009

7. 0.68   8. 0.12   9. 0.107

Write as a percent. Round to the nearest tenth of a percent.

10. $\dfrac{25}{60}$ 11. $1\dfrac{8}{9}$ 12. $\dfrac{4}{7}$

13. $\dfrac{5}{32}\%$ 14. $\dfrac{3}{5}$ 15. $1\dfrac{5}{6}$

Write as a percent. Write the remainder in fractional form.

16. $\dfrac{5}{11}$ 17. $\dfrac{2}{9}$ 18. $1\dfrac{1}{7}$

19. $\dfrac{3}{8}$ 20. $\dfrac{1}{15}$ 21. $\dfrac{7}{12}$

1. _____
2. _____
3. _____
4. _____
5. _____
6. _____
7. _____
8. _____
9. _____
10. _____
11. _____
12. _____
13. _____
14. _____
15. _____
16. _____
17. _____
18. _____
19. _____
20. _____
21. _____

Name   Score

Solve.

1. 8% of 45 is what?   2. 15% of 100 is what?   1. _____

                                                                         2. _____

3. 26% of 70 is what?   4. 53% of 90 is what?   3. _____

                                                                         4. _____

5. What is 45% of 60?   6. What is 35% of 55?   5. _____

                                                                         6. _____

7. What is 65% of 135.5?   8. What is 52% of 24.4?   7. _____

                                                                         8. _____

9. What is 4% of 2800?   10. What is 0.02% of 250?   9. _____

                                                                         10. _____

11. 5% of 900 is what?   12. 0.025% of 800 is what?   11. _____

                                                                         12. _____

13. 150% of 98 is what?   14. 220% of 6 is what?   13. _____

                                                                         14. _____

15. Find 12% of 540.   16. Find 30% of 17.5.   15. _____

                                                                         16. _____

17. Find 7.9% of 120.   18. Find 2.5% of 440.   17. _____

                                                                         18. _____

19. Find 0.08% of 375   20. Find 25% of 144.   19. _____

                                                                         20. _____

21. What is 12.5% of 1540?   22. What is $6\frac{1}{4}$% of 620? 21. _____

                                                                         22. _____

Name   Score

Solve.

1. 34% of 850 is what?   2. What is 15% of 400?   1. _____

                                                                        2. _____

3. 54 is what percent of 180?   4. What percent of 120 is 42?   3. _____

                                                                        4. _____

5. 90% of what is 54?   6. 146 is 73% of what?   5. _____

                                                                        6. _____

7. What percent of 324 is 162?   8. What percent of 140 is 35?   7. _____

                                                                        8. _____

9. 153 is 30% of what?   10. 7.5% of what is 60?   9. _____

                                                                        10. _____

11. 222 is what percent of 600?   12. 144 is what percent of 32?   11. _____

                                                                        12. _____

13. What is 160% of 480?   14. 210% of 390 is what?   13. _____

                                                                        14. _____

15. 27 is 45% of what?   16. 24% of what is 36?   15. _____

                                                                        16. _____

17. What percent of 110 is 44?   18. 65 is what percent of 125?   17. _____

                                                                        18. _____

19. What percent of 220 is 11?   20. What percent of 72 is 9?   19. _____

                                                                        20. _____

21. 91 is what percent of 26?   22. What percent of 46 is 184?   21. _____

                                                                        22. _____

23. 0.2% of what is 33?   24. 15 is 0.6% of what?   23. _____

                                                                        24. _____

Name _____   Score _____

Solve.

1. A nursery sold 450 geranium plants in June. In July, the nursery increased its sales by 8%. How many geranium plants were sold in July?

2. A restaurant serves 35% more ice cream in June than it does in May. How much ice cream does the restaurant serve in June if it serves 12 gallons of ice cream in May?

1. _____

2. _____

3. A truck retail sales company makes a 5.3% profit on sales of $520,000. Find the profit.

4. An office building has an appraised value of $8,000,000. The real estate taxes are 2.35% of the appraised value of the building. Find the real estate taxes.

3. _____

4. _____

5. A used car salesperson sold 6 of the 50 cars in the lot. What percent of the total number of cars in the lot were sold?

6. An investor received a dividend of $540 on an investment of $4500. What percent of the investment is the dividend?

5. _____

6. _____

7. A survey of 1760 people showed that 352 people favored the incumbent mayor. What percent of the people surveyed favored the incumbent mayor?

8. Of the 3900 resistors tested, 117 were found defective. What percent of the total number of resistors were defective?

7. _____

8. _____

9. A total of $4200 was paid in taxes on an income of $16,800. Find the percent of the total income paid in taxes.

10. An advertising survey of 265 people found that 53 liked a new toothpaste. What percent of the people surveyed did not like the new toothpaste?

9. _____

10. _____

11. A soccer team won 51 out of 68 games they played. What percent of the games played did they win?

12. During shipping, 600 of the 4800 light bulbs were damaged. What percent of the number of light bulbs were damaged in the shipping?

11. _____

12. _____

13. During an inspection of 120 motorcycles, 3 of them did not pass the safety test. What percent of the motorcycles inspected did pass the safety test?

14. An administrative assistant types 70 words per minute with 98% accuracy. During 5 minutes of typing, how many errors does the secretary make?

13. _____

14. _____

Name _____   Score _____

Solve. Round to the nearest tenth of a percent or to the nearest cent.

1. The value of a $7000 investment increased $1750. What percent increase does this represent?

2. A sweater which sold for $24 last month increased in price by $2. What percent increase does this represent?

1. _____

2. _____

3. The average price of fuel oil rose from $0.75 to $1.00 in six months. What was the percent increase in the price of fuel oil?

4. The number of students enrolled in a speed reading course increased from 60 to 66 during the first 10 days of school. What is the percent increase?

3. _____

4. _____

5. A manufacturer of ceiling fans increased its monthly output of 1500 by 10%. What is the amount of increase?

6. An advertising agency increased its 200 billboards by 15% during the past year. How many billboards does the agency have now?

5. _____

6. _____

7. The employees of a manufacturing plant received a 6% increase in pay.
   a. What is the amount of the increase for an employee who makes $225 per week?
   b. What is the weekly wage for the employee after the wage increase?

8. A supervisor's salary this year is $32,000. This salary will increase by 8% next year.
   a. What is the amount of increase?
   b. What will the salary be next year?

7. a. _____

   b. _____

8. a. _____

   b. _____

9. A college increased its number of parking spaces from 1000 to 1050.
   a. How many new spaces were added?
   b. What percent increase does this represent?

10. A town plans to increase its 4000 water meters by 7.5%.
    a. How many more water meters is this?
    b. What will the total number of water meters be after this increase?

9. a. _____

   b. _____

10. a. _____

    b. _____

11. A cafeteria increased the number of items on the menu from 80 to 90. What percent of increase does this represent?

12. The amount of gasoline used by a fleet of cars increased from 200 to 230 gallons per day. What percent increase does this represent?

11. _____

12. _____

Name _____   Score _____

Solve.

1. A health spa sold 120 memberships in November. In May the spa sold 18 fewer memberships than in November. What was the percent decrease in the number of memberships sold?

2. A new bypass around a small town reduced the normal 40-minute driving time between two cities by 12 minutes. What percent decrease does this represent?

3. By washing all the clothes in cold water, a family was able to reduce its normal monthly utility bill of $125 by $15. What percent decrease does this represent?

4. By installing solar panels, a copying center reduced its normal $240 per month heating bill by $36. What percent decrease does this represent?

5. A golf resort employs 180 people during the golfing season. At the end of the season, the resort reduces the number of employees by 45%. What is the decrease in the number of employees?

6. It is estimated that the value of a motorcycle is reduced by 25% after one year of ownership. Using this estimate, how much value does a $1500 new motorcycle lose after one year?

7. A new process reduced the time needed to replate a piece of silverware from 16 minutes to 10 minutes.
   a. What is the amount of decrease?
   b. What percent decrease does this represent?

8. Because of a decrease in orders for telephones, a telephone center reduced the orders for phones from 140 per month to 91 per month.
   a. What is the amount of decrease?
   b. What percent decrease does this represent?

9. Last year a company earned a profit of $285,000. This year, the company's profits were 6% less than last year's.
   a. What was the amount of decrease?
   b. What was the profit this year?

10. Last year, a laptop computer cost $420 to produce. Mass production enabled this manufacturer to reduce this expense by 15%.
    a. What is the amount of decrease?
    b. What was the cost after the reduction?

11. The price of a new model digital camera dropped from $150 to $114 in ten months. What percent decrease does this represent?

12. As a result of computerized cash registers, the average customer check-out time has decreased from 10 minutes to 9.5 minutes. What percent decrease does this represent?

1. _____

2. _____

3. _____

4. _____

5. _____

6. _____

7. a. _____

   b. _____

8. a. _____

   b. _____

9. a. _____

   b. _____

10. a. _____

    b. _____

11. _____

12. _____

Name _____ Score _____

Solve.

1. A store manager used a markup rate of 30% on all desk lamps. What is the markup on a lamp which costs the store $26?

2. A manager of a natural food store determines that the markup rate of 17% is necessary to make a profit. What is the markup on an item which costs the dealer $2?

3. An automobile tire dealer uses a markup rate of 32%. What is the markup on tires which cost the dealer $34?

4. The markup on a necklace which costs a jeweler $60 is $36. What markup rate does this represent? (Solve the basic percent equation for percent.)

5. A beach-wear shop uses a markup rate of 40% on a bathing suit which costs the shop $34.
    a. What is the markup?
    b. What is the selling price?

6. A department store uses a markup rate of 44% on its Model XL food processor which costs the store $45.
    a. What is the markup?
    b. What is the selling price?

7. A produce market pays $1.09 for pineapples. The market uses a 55% markup rate.
    a. What is the markup?
    b. What is the selling price?

8. A garden shop uses a markup rate of 35% on a rose trellis which costs the store $26.
    a. What is the markup?
    b. What is the selling price?

9. A store uses a markup rate of 38%. What is the selling price for the DVD player which costs the store $47?

10. What is the selling price on a pair of jogging shoes which costs a store $31? The store uses a markup rate of 44%.

1. _____

2. _____

3. _____

4. _____

5. a. _____

   b. _____

6. a. _____

   b. _____

7. a. _____

   b. _____

8. a. _____

   b. _____

9. _____

10. _____

Name   Score

Solve.

1. To promote business, a store manager offers a small vacuum cleaner which regularly sells for $25 at $9 off the regular price. What is the discount rate?

2. A department store is giving a discount of $6 on an ice chest which normally sells for $40. What is the discount rate?

3. A sporting goods store is selling its $150 exercise bike for 20% off the regular price. What is the discount?

4. A jewelry store is selling $250 quartz watches at 30% off the regular price. What is the discount?

5. A stereo speaker set which regularly costs $450 is on sale for $90 off the regular price. What is the discount rate?

6. A hardware store is selling its $32 lock set for 15% off the regular price. What is the discount?

7. Colby cheese which regularly sells for $2.40 per pound is on sale for 25% off the regular price.
   a. What is the discount?
   b. What is the sale price?

8. The pro shop at the racketball club has its regularly priced $55 shoes on sale for 18% off the regular price.
   a. What is the discount?
   b. What is the sale price?

9. A gift shop has its picture frames which regularly cost $35 on sale for $30.80.
   a. What is the discount?
   b. What is the discount rate?

10. An automobile body shop has regularly priced $600 paint jobs on sale for $480.
   a. What is the discount?
   b. What is the discount rate?

11. During a going-out-of-business sale, all lawn and garden merchandise was reduced 40% off the regular price. What was the sale price of a lawn mower which normally sells for $230?

12. A store offering 35% off its stock of art supplies. What is the sale price of a set of paint brushes which regularly sells for $60?

1. _____

2. _____

3. _____

4. _____

5. _____

6. _____

7. a. _____

   b. _____

8. a. _____

   b. _____

9. a. _____

   b. _____

10. a._____

    b. _____

11. _____

12. _____

Name _____ Score _____

Solve.

1. A rancher borrows $120,000 for 10 months at an annual interest rate of 18%. What is the simple interest due on the loan?

2. To finance the purchase of 8 new taxicabs, the owners of the fleet borrows $84,000 for 8 months at an annual interest rate of 16%. What is the simple interest due on the loan?

3. A mobile home dealer borrowed $160,000 at a 15.5% annual interest rate for four years. What is the simple interest due on the loan?

4. An executive was offered a $34,000 loan at a 14.5% annual interest rate for three years. Find the simple interest due on the loan.

5. You arrange for a 6-month bank loan of $6000 at an annual simple interest rate of 7.5%. Find the total amount you must repay to the bank.

6. A bank charges its customers an interest rate of 1.8% per month for transferring money into an account which is overdrawn. Find the interest owed to the bank for one month when $300 was transferred into an overdrawn account.

7. A copier is purchased and a $2100 loan is obtained for two years at a simple interest rate of 17%.
    a. Find the interest due on the loan.
    b. Find the maturity value of the loan.

8. An oil well drilling company purchased two new helicopters for $65,000 and financed the full amount at 9% simple annual interest for four years.
    a. Find the interest due on the loan.
    b. Find the maturity value of the loan.

9. A company purchased a computer system for $75,000 and financed the full amount for five years at a simple annual interest rate of 15%.
    a. Find the interest due on the loan.
    b. Find the maturity value of the loan.

10. To reduce their inventory of new cars, a dealer is offering car loans at a simple annual interest rate of 11%.
    a. Find the interest charged to a customer who financed a car loan of $8200 for four years.
    b. Find the maturity value of the loan.

1. _____

2. _____

3. _____

4. _____

5. _____

6. _____

7. a. _____

   b. _____

8. a. _____

   b. _____

9. a. _____

   b. _____

10. a. _____

    b. _____

Name   Score

Solve.

1. How many degrees are in one half of a revolution?

2. Find the supplement of a 116° angle.

1. _____

2. _____

3. Find the complement of a 75° angle.

4. Find the complement of a 22° angle.

3. _____

4. _____

5. Find the supplement of a 128° angle.

6. Find the complement of a 49° angle.

5. _____

6. _____

7. Find the complement of a 66° angle.

8. Find the supplement of a 138° angle.

7. _____

8. _____

9. Find the supplement of a 42° angle.

10. Find the complement of a 12° angle.

9. _____

10. _____

11. Find the complement of a 81° angle.

12. Find the supplement of a 171° angle.

11. _____

12. _____

13. Angle AOB is a straight line. Find $\angle AOC$.

14. Angle AOB is a straight line. Find $\angle COB$.

13. _____

14. _____

15. Find $\angle x$.

16. Find $\angle AOB$.

15. _____

16. _____

17. Find $\angle AOC$.

18. Find $\angle A$.

17. _____

18. _____

Name _____ Score _____

Find *x*.

1.

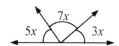

2.

1. _____

3.

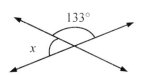

4.

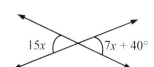

2. _____

3. _____

4. _____

Find the measures of angles *a* and *b*. $k \parallel \ell$

5.

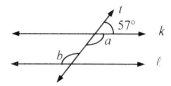

6.

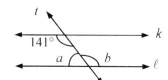

5. _____

6. _____

Find *x*. $k \parallel \ell$

7.

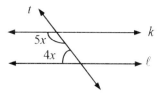

8.

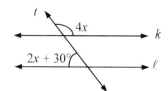

7. _____

8. _____

Name   Score

Solve.

1. A triangle has a 40° angle and a right angle. What is the measure of the third angle?

2. A triangle has a 53° angle and a right angle. Find the measure of the third angle.

1. _____

2. _____

3. Two angles of a triangle measure 18° and 107°. Find the measure of the third angle.

4. Two angles of a triangle measure 68° and 43°. Find the measure of the third angle.

3. _____

4. _____

5. A triangle has a 121° angle and a 45° angle. What is the measure of the third angle?

6. A triangle has a 79° angle and a 72° angle. What is the measure of the third angle?

5. _____

6. _____

7. Two angles of a triangle measure 92° and 58°. What is the measure of the third angle?

8. A triangle has a 84° angle and a 39° angle. Find the measure of the third angle.

7. _____

8. _____

Name _____   Score _____

Simplify.

1. Find the perimeter of a triangle in which each side is 18 cm.

2. Find the perimeter of a triangle with sides 21.3 cm, 17.4 cm, and 14.8 cm.

3. Find the circumference of a circle with radius of 6 cm. Use 3.14 for π.

4. Find the circumference of a circle with a diameter of 28 in. Use $\frac{22}{7}$ for π.

5. Find the perimeter of a rectangle with a width of 5 ft and a length of $8\frac{1}{2}$ ft.

6. Find the perimeter of a square in which the sides are equal to $8\frac{1}{4}$ ft.

7. Find the perimeter of a square in which the sides are equal to 15.5 m.

8. Find the perimeter of a triangle with sides 3 ft 8 in., 5 ft 6 in., and 6 ft 9 in.

9. Find the perimeter of a five-sided figure with sides of 19 cm, 36 cm, 25 cm, 39 cm, and 20 cm.

10. Find the perimeter of a six-sided figure with sides of 15 ft, 32 ft, 21 ft, 34 ft, 17 ft, and 26 ft.

Solve.

11. Find the number of feet of fence needed to fence a park that is $1\frac{3}{8}$ mi long and $\frac{5}{8}$ mi wide.

12. A horse trainer is planning to enclose a circular exercise area. How much fencing is needed if it has a diameter of 80 ft. Use 3.14 for π.

13. An irrigation system waters a circular field which has a 60-foot radius. Find the distance around the outside edge of the field. Use 3.14 for π.

14. Find the length of weather stripping needed to put around a rectangular window that is 9 ft high and 22 ft long.

15. Find the length of aluminum framing needed to frame a picture that is 6 ft by 4 ft.

16. Find the length of a rubber gasket needed to fit around a circular porthole that has a 16 in. diameter. Use 3.14 for π.

1. _____

2. _____

3. _____

4. _____

5. _____

6. _____

7. _____

8. _____

9. _____

10. _____

11. _____

12. _____

13. _____

14. _____

15. _____

16. _____

Name   Score

Solve.

1. Find the area of a triangle with a base of 8 ft and a height of $1\frac{3}{4}$ ft.

2. Find the area of a right triangle with a base of 11 cm and a height of 6.4 cm.

1. _____

2. _____

3. Find the area of a square with a side of 7 ft.

4. Find the area of a square with a side of 29 cm.

3. _____

4. _____

5. Find the area of a rectangle with a length of 52 cm and a width of 27 cm.

6. Find the area of a rectangle with a length of 25 in., and a width of 13 in.

5. _____

6. _____

7. Find the area of a circle with a radius of 11 in. Use 3.14 for π.

8. Find the area of a square with a side of 7.8 cm.

7. _____

8. _____

9. Find the area of a triangle with a base of 27 cm and a height of 30 cm.

10. Find the area of a rectangle with a length of 39 cm and a width of 25 cm.

9. _____

10. _____

11. Find the area of a circular rug that is 18 ft in diameter.

12. Find the area of a circular portrait that has a 24 in. diameter.

11. _____

12. _____

13. A circular skating rink has a 144-foot radius. Find the area of the surface of the ice rink.

14. Find the area of a rectangular park that is 300 m long and 180 m wide.

13. _____

14. _____

Name _____   Score _____

Find the unknown side of the triangle. Round to the nearest thousandth.

1.
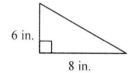
6 in.
8 in.

2.

14 cm
12 cm

1. _____

2. _____

3.

15 ft
12 ft

4.

18 in.
16 in.

3. _____

4. _____

5.

10 ft
8 ft

6.

10 yd
10 yd

5. _____

6. _____

Solve. Round to the nearest thousandth.

7. A guy wire holds a television antenna in place. Find the distance along the ground from the base of the antenna to the guy wire.

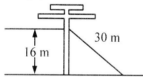
30 m
16 m

8. Find the length of a ramp used to move a fork lift from the ground to the loading ramp.

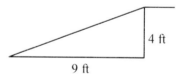
4 ft
9 ft

7. _____

8. _____

9. A car is driven 24 mi west and then 10 mi north. How far is the car from the starting point?

10 mi
24 mi

10. The sketch below shows a map of a hiking trail. Find the distance around the hiking trail.

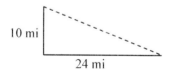
6 km
8 km

9. _____

10. _____

Name   Score _____

Find the ratio of the corresponding sides for the similar triangles.

1. 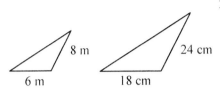   2.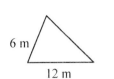

1. _____

2. _____

Triangles ABC and DEF and similar. Find the indicated distance. Round to the nearest tenth.

3.   Find side AC.   4.   Find side BC.   3. _____

   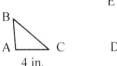   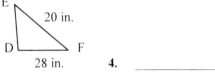

4. _____

5.   Find the height of triangle ABC.   6.   Find the height of triangle DEF.   5. _____

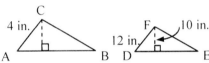   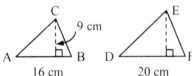

6. _____

Solve.

7.   Find the height of the flagpole.   8.   Find the height of the building.   7. _____

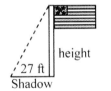   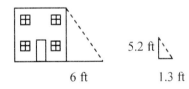

8. _____

Triangles ABC and DEF are similar.

9.   Find the perimeter of triangle ABC. 10. Find the area of triangle DEF.   9. _____

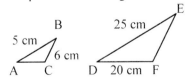   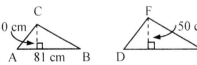

10. _____

Name _____   Score _____

Determine whether the two triangles are congruent. If they are congruent, state by what rule they are congruent.

1. 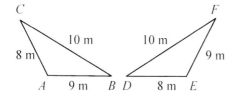 2.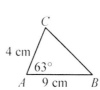

1. _____

2. _____

3.   4.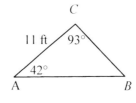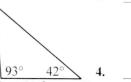

3. _____

4. _____

5. Given triangle *ABC* and triangle *DEF*, do the conditions $BC = EF$, $AC = DF$ and $\angle C = \angle F$ guarantee that triangle *ABC* is congruent to triangle *DEF*? If they are congruent, by what rule are they congruent?

6. Given triangle *LMN* and triangle *PQR*, do the conditions $\angle L = \angle P$, $\angle M = \angle Q$ and $LM = PQ$ guarantee that triangle *LMN* is congruent to triangle *PQR*? If they are congruent, by what rule are they congruent?

5. _____

6. _____

7. Given triangle *DEF* and triangle *JKL*, do the conditions $DE = JK$, $EF = KL$, and $DF = JL$ guarantee that triangle *DEF* is congruent to triangle *JKL*? If they are congruent, by what rule are they congruent?

8. Given triangle *QRS* and triangle *JKL*, do the conditions $\angle S = \angle L$, $QR = JK$, and $RS = KL$ guarantee that triangle *QRS* is congruent to triangle *JKL*? If they are congruent, by what rule are they congruent?

7. _____

8. _____

Name _____   Score _____

Solve.

1. Find the volume of a cylinder with a radius of 10 cm and a height of 21 cm. Use $\frac{22}{7}$ for π.

2. Find the volume of a cylinder with a radius of 22 cm and a height of 35 cm. Use $\frac{22}{7}$ for π.

1. _____

2. _____

3. Find the volume of a cube with a side of 5 ft 3 in. Round to the nearest hundredth.

4. Find the volume of a cube with a side of 22 cm.

3. _____

4. _____

5. Find the volume of a cube with a side of 6 ft 4 in. Round to the nearest hundredth.

6. Find the volume of a sphere with a radius of 6 mm. Use 3.14 for π. Round to the nearest hundredth.

5. _____

6. _____

7. Find the volume of a sphere with a radius of 12 cm. Use 3.14 for π. Round to the nearest hundredth.

8. Find the volume of a sphere with a radius of 1.4 mm. Use 3.14 for π. Round to the nearest hundredth.

7. _____

8. _____

9. Find the volume of a rectangular solid with a length of 2 m 40 cm, and a width of 250 cm, and a height of 191 cm.

10. Find the volume of a rectangular solid with a length of 5 m, and width of 240 cm, and a height of 150 cm.

9. _____

10. _____

11. A propane gas storage tank which is in the shape of a cylinder, is 10 m high and has a 6-meter diameter. Find the volume of the gas storage tank. Use 3.14 for π.

12. A silo, which is the shape of a cylinder, is 27 ft in diameter and has a height of 48 ft. Find the volume of the silo. Use 3.14 for π.

11. _____

12. _____

13. Find the volume of a spherical oxygen tank that is 9 m in diameter. Use 3.14 for π. Round to the nearest tenth.

14. Find the volume of a spherical water tank that is 18 ft in diameter. Use 3.14 for π.

13. _____

14. _____

15. Find the volume of a railroad car that is 16 m long, 5 m wide, and 5.2 m high.

16. A hole is being dug for installing a swimming pool. The hole is 24 ft long, 12 ft wide, and 8 ft deep. Find the volume of the hole.

15. _____

16. _____

Name _____   Score _____

Solve.

1. The side of a cube measures 5.3 cm.
    Find the surface area of the cube.

2. Find the surface area of a sphere
    with radius 7 m. Round to the nearest
    hundredth.

1. _____

2. _____

3. The diameter of the base of a cylinder
    is 10 in. The height is 15 in. Find the
    surface area. Given the exact value.

4. The length of a side of the base of a
    regular pyramid is 20 m and the slant
    height is 25 m. Find the surface area
    of the pyramid.

3. _____

4. _____

5. The surface area of a rectangular solid
    is 423 square feet. The length of the
    solid is 11 ft and the width is 8 ft.
    Find the height of the rectangular solid.

6. The slant height of a cone is 4.5 in.
    The radius of the base is 3.5 in. Find
    the surface area. Give the exact value.

5. _____

6. _____

7. A can of paint will cover 300 ft^2. How
    many cans of paint should be
    purchased in order to paint a cylinder
    that has a height of 20 ft and a radius
    of 8 ft.

8. Find the area of a label used to cover
    a can that has a diameter of 12.5 cm
    and a height of 15 cm. Round to the
    nearest hundredth.

7. _____

8. _____

Name _____   Score _____

Use the table below.

Number of hours worked at a part-time job per week for 40 students

10	21	16	8	12	10	24	2
26	6	11	18	31	3	14	22
11	7	5	32	27	30	4	10
14	16	23	3	8	18	21	13
27	10	1	22	18	25	30	12

1. What is the range of the data in the table?

2. Make a frequency distribution table for the number of hour worked. Use 8 classes.

1. _____

2. _____

3. Which class has the greatest frequency?

4. How many students worked between 22 and 32 hours?

3. _____

4. _____

5. How many students worked between 9 and 16 hours?

6. What percent of students worked between 5 and 12 hours?

5. _____

6. _____

7. What percent of students worked between 17 and 21 hours?

8. How many students worked 21 hours or less?

7. _____

8. _____

Name   Score

The fuel usage of 100 cars was measured by a research group. The results are recorded in the histogram below.

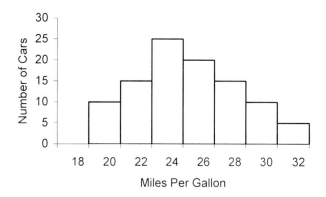

1. Find the number of cars which get between 24 and 26 miles per gallon.

2. Find the number of cars which get between 30 and 32 miles per gallon.

3. Find the number of cars which get 24 or more miles per gallon.

4. What percent of the number of cars which get between 18 and 22 miles per gallon?

1. _____

2. _____

3. _____

4. _____

The hourly wages of the 100 employees of a company are recorded in the histogram below.

5. Find the number of employees whose hourly wage is between $6 and $10.

6. Find the ratio of the number of employees whose hourly wage is between $12 and $14 to the total number of employees.

7. Find the number of employees whose hourly wages is between $8 and $14.

8. How many employees earn $12 or more per hour?

5. _____

6. _____

7. _____

8. _____

Name   Score

A radio rating service surveyed 105 families to find the number of hours they listened to the radio. The results are recorded in the figure below.

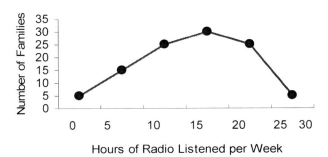

Hours of Radio Listened per Week

1. How many families listened between 10 and 15 hours a week?

2. What is the ratio of the number of families who listened between 20 and 25 hours a week to the total number in the survey?

1. _____

2. _____

3. How many families listened between 20 and 30 hours a week?

4. What is the ratio of the number of families who listened 15 or more hours a week to the total number in the survey?

3. _____

4. _____

A real estate company sold 100 homes during the last three months. The selling prices are recorded in the figure below.

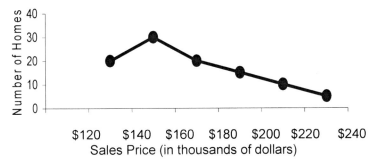

Sales Price (in thousands of dollars)

5. How many homes sold for between $140,000 and $160,000?

6. Find the ratio of the number of homes which sold for between $120,000 and $140,000 to the total number of homes sold during the three months.

5. _____

6. _____

7. How many homes sold for between $160,000 and $240,000?

8. How many homes sold for between $120,000 and $180,000?

7. _____

8. _____

Name   Score

Solve.

1. The prices of a scientific calculator at five stores were $16.25, $15.75, $16.15, $16.50, and $16.95. Find the mean price of the calculator.

2. A student received grades of 83, 89, 85, 91, and 92 on five mathematics exams. Find the mean grade of the student's mathematics exams.

1. _____

2. _____

3. The six sales representatives for an advertising agency received weekly bonuses of $345, $275, $190, $221, $335, and $260. Find the mean bonus.

4. A number of pizzas sold in five different stores over a three-day period was 212, 246, 205, 252, and 245. Find the mean number of pizzas sold per store during this period of time.

3. _____

4. _____

5. The prices of identical CD clock radios at each of five stores were $33.25, $39.00, $36.75, $36.00, and $37.50. Find the median price of the CD clock radio.

6. The number of miles driven during each of five days of a business trip was 107, 96, 151, 103, and 99. Find the median number of miles driven.

5. _____

6. _____

7. The hourly wages for seven job classifications at a company are $7.63, $10.43, $10.09, $7.59, $8.45, $7.47, and $8.38. Find the median hourly wage.

8. The ages of the seven most recently hired employees at a fast food restaurant are 22, 41, 19, 21, 20, 26, and 29. Find the median age.

7. _____

8. _____

9. The number of responses to a discount coupon for a carpet cleaner during a six-day period was 35, 29, 33, 37, 28 and 26. What is the mode of the data?

10. The scores on eight math exams at a job placement service were 79, 93, 95, 69, 93, 96, 88, and 75. What is the mode of the data?

9. _____

10. _____

Name   Score

The ages of the 300 accountants who passed the certified public accountant (CPA) exam at one test center were recorded. The box-and-whiskers plot below shows the distribution of their scores.

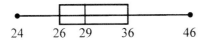

24   26   29   36   46

1.   What is the youngest age?   2.   What is the oldest age?   1. _____

                                                                                          2. _____

3.   What is the first quartile?   4.   What is the third quartile?   3. _____

                                                                                          4. _____

5.   What is the median?   6.   What is the range?   5. _____

                                                                                          6. _____

7.   What is the interquartile range?   8.   How many of the accountants were   7. _____
                                                    older than 36?

                                                                                          8. _____

9.   How many of the accountants were   10. How many accountants are represented 9. _____
     younger than 29?   in each quartile?

                                                                                          10. _____

11. What percent of the accountants were   12. What percent of the accountants were 11. _____
     younger than 36?   older than 46?

                                                                                          12. _____

Name _____   Score _____

Solve.

1.   A student received grades of 84, 90, 83, 76 and 92 on five exams.   1. _____
      Find the standard deviation of these exams.

2.   The ages of six students are 18, 21, 22, 19, 19, and 21.   2. _____
      Find the standard deviation of these ages.

3.   Seven coins were tossed 20 times. The number of heads recorded were 8,   3. _____
      12, 11, 9, 12, 9, and 9. Find the standard deviation of the number of heads.

4.   The scores for five professional golfers were 68, 70, 75, 69 and 73. The scores   4. _____
      for five amateur golfers were 84, 81, 89, 86, and 85. Which scores had the
      greater standard deviation?

Name _____   Score _____

Solve.

1. Two dice are rolled. What is the probability that the sum of the dots on the upwards faces is 4?

2. Two dice are rolled. What is the probability that the sum of the dots on the upwards faces is 9?

1. _____

2. _____

3. Two dice are rolled. What is the probability that the sum of the dots on the upwards faces is greater than 1?

4. A coin is tossed 3 times. What is the probability that the outcomes of the tosses consist of two tails and one head?

3. _____

4. _____

5. Each of the letters of the word *MISSISSIPPI* is written on a card, and the cards are placed in a hat. One card is drawn at random from the hat. What is the probability that the card has the letter *S* on it?

6. Each of the letters of the word *MISSISSIPPI* is written on a card, and the cards are placed in a hat. One card is drawn at random from the hat. What is the probability that the card has the letter *P* on it?

5. _____

6. _____

7. Which has a greater probability, drawing a 5, 6, or 10 from a deck of cards or drawing a diamond?

8. In a psychology class, a set of exams earned the following grades: 6 A's, 9 B's, 16 C's, 5 D's, and 3 F's. If a single student's exam is chosen from this class, what is the probability that it received an A?

7. _____

8. _____

9. Six purple marbles, four orange marbles, and eight blue marbles are placed in a bag. One marble is chosen at random. What is the probability that the marble chosen is purple?

10. Seven purple marbles, five orange marbles, and three blue marbles are placed in a bag. One marble is chosen at random. What is the probability that the marble chosen is orange?

9. _____

10. _____

Name _____   Score _____

Solve.

1. A coin is tossed twice. What are the odds of its showing heads both times?

2. Two dice are rolled. What are the odds in favor of rolling a 2?

    1. _____

    2. _____

3. Two dice are rolled. What are the odds in favor of rolling a 6?

4. A single card is selected from a standard deck of playing cards. What are the odds against its being a jack?

    3. _____

    4. _____

5. A single card is selected from a standard deck of playing cards. What are the odds against its being a spade?

6. The odds in favor of a candidate winning an election is 4 to 3. What is the probability of the candidate winning the election?

    5. _____

    6. _____

7. Two dice are rolled. What are the odds against rolling a 3?

8. At the beginning of the professional football season, one team was given 20 to 1 odds against its winning the Super Bowl. What is the probability of this team winning the Super Bowl?

    7. _____

    8. _____

CHAPTER 1

Objective 1.1A

1.
2.
3.
4.

5. >   6. >   7. <   8. <   9. >   10. <   11. <   12. >   13. <   14. >   15. >   16. <
17. 6, 8, 21, 57, 80, 81   18. 6, 14, 19, 64, 72   19. 200, 312, 435, 812, 814

Objective 1.1B

1. Eight hundred sixty-two   2. Three hundred eight   3. Five thousand one hundred twenty-five
4. Three hundred eighty thousand seven hundred fifty-one   5. Eight hundred thousand one
6. Seven million six hundred forty thousand seven hundred twenty-three   7. 33   8. 274
9. 9527   10. 56,320   11. 460,303   12. 4,012,986   13. 1,000,0005   14. 8,001, 050
15. 200 + 50 + 6   16. 4000 + 700 + 3   17. 50,000 + 900 + 10 + 6   18. 500,000 + 400
19. 900,000 + 20,000 + 10   20. 4,000,000 + 200,000 + 70,000 + 1000 + 20

Objective 1.1C

1. 750   2. 710   3. 400   4. 600   5. 1100   6. 7100   7. 4000   8. 10,000
9. 75,000   10. 69,000   11. 250,000   12. 840,000   13. 3,000,000   14. 8,000,000

Objective 1.1D

1. yours   2. Veggie Burger   3. 100 watt bulb   4. regular mattress   5. Montana
6. middle   7. 2200 miles   8. 9000 square miles   9. $14,000,000,000   10. Denmark

Objective 1.2A

1. 866   2. 852   3. 2595   4. 18,340   5. 20,631   6. 19,630   7. 1439   8. 1271
9. 74,748   10. 34,430   11. 1096   12. 77,612   13. 10,130   14. 1318
15. 9498   16. 76,017   17. No   18. No   19. Yes

Objective 1.2B

1. 95   2. 400   3. 71   4. 8413   5. 817   6. 2204   7. 3269   8. 4399
9. 966   10. 3919   11. 26,699 12. 43,848   13. 19,027   14. 151,271   15. 353,874   16. 7006
17. 421   18. 2111   19. 6151   20. 24,789   21. 56,741   22. Yes   23. No   24. Yes

Objective 1.2C

1. $81   2. $340   3. $211   4. $650   5. $385   6. 169,064 square miles   7. 3235 people
8. 239 people   9. $1554   10. 62,363 miles   11. $23   12. $7724

Objective 1.3A

1. 392   2. 1860   3. 3150   4. 17,984   5. 18,315   6. 47,320   7. 181,815
8. 601,644   9. 198,170   10. 864,000   11. 1,154,160   12. 2,998,128   13. 304   14. 72
15. 6372   16. 150   17. Yes   18. No   19. No   20. Yes

Objective 1.3B

1. 4^4 **2.** 8^6 **3.** $3^4 \cdot 4^2$ **4.** $5^4 \cdot 7^3$ **5.** $2^3 \cdot 3^4 \cdot 7^2$ **6.** $5 \cdot 6 \cdot 7^2 \cdot 8^3$ **7.** $4^2 \cdot 5^3 \cdot 6^3 \cdot 7$
8. $7^3 \cdot 15^2 \cdot 19^2$ **9.** 16   **10.** 27   **11.** 64   **12.** 0   **13.** 320   **14.** 37,500
15. 102,900   **16.** 13,824   **17.** 39,690   **18.** 3888   **19.** 2000   **20.** 25,088   **21.** 139,968
22. 0   **23.** 14,400

Objective 1.3C

7. 18 r 1   **8.** 6 r 6   **9.** 4 r 4   **4.** 247   **5.** 621   **6.** 57   **7.** 173 r 1
8. 50 r 5   **9.** 148 r 2   **10.** 8217 r 1   **11.** 14,850 r 3   **12.** 9080 r 1   **13.** 7   **14.** Undefined
15. 2537   **16.** 13   **17.** Yes   **18.** Yes   **19.** No   **20.** No

Objective 1.3D

1. 1, 3, 5, 15   **2.** 1, 23   **3.** 1, 2, 4, 5, 8, 10, 20, 40   **4.** 1, 2, 31, 62   **5.** 1, 3, 9, 27, 81
6. 1, 3, 17, 51   **7.** 1, 2, 4, 8, 11, 22, 44, 88   **8.** 1, 3, 7, 9, 21, 63   **9.** 1, 5, 7, 35   **10.** 1, 3, 5, 9, 15, 45
11. 1, 2, 3, 6, 13, 26, 39, 78   **12.** 1, 5, 11, 55   **13.** $2 \cdot 2 \cdot 2$ **14.** $2 \cdot 3 \cdot 5$ **15.** $1 \cdot 53$ **16.** $1 \cdot 11$
17. $2 \cdot 2 \cdot 5$ **18.** $2 \cdot 41$ **19.** $5 \cdot 7$ **20.** $2 \cdot 2 \cdot 11$ **21.** $3 \cdot 23$ **22.** $2 \cdot 2 \cdot 2 \cdot 3 \cdot 3$
23. $2 \cdot 2 \cdot 2 \cdot 11$ **24.** $2 \cdot 47$

Objective 1.3E

1. 228 miles   **2.** 318 miles   **3.** $184,250   **4.** 160 people   **5.** 864 books   **6.** $9456   **7.** $212
8. $39   **9.** 8508 hours   **10.** $630   **11.** $115   **12.** $184

Objective 1.4A

1. $x = 6$ **2.** $n = 3$ **6.** $x = 0$ **7.** $t = 2$ **8.** $v = 4$ **9.** $x = 3$ **10.** $x = 1$
13. $y = 10$ **19.** $z = 2$ **10.** 44   **11.** 32   **12.** 13   **1.** $x = 4$ **2.** $x = 5$
5. $y = 6$ **7.** $y = 4$ **17.** 1   **18.** 2   **19.** 3   **20.** 5   **21.** 4

Objective 1.4B

1. 5   **2.** 20   **3.** 13 in.   **4.** 9 hrs   **5.** 6 hrs   **6.** 27   **7.** 29
8. 48 payments   **9.** 30 months   **10.** 8 in.

Objective 1.5A

1. 10   **2.** 11   **3.** 19   **4.** 0   **5.** 8   **6.** 5   **7.** 6   **8.** 11   **9.** 8   **10.** 10   **11.** 79   **12.** 0   **13.** 23   **14.** 2
15. 22   **16.** 16   **17.** 84   **18.** 7   **19.** 2   **20.** 38   **21.** 13   **22.** 8   **23.** 33   **24.** 4   **25.** 9   **26.** 5   **27.** 10

CHAPTER 2

Objective 2.1A

1.   **2.**
3.   **4.**
5. 2   **6.** −1   **7.** 4   **8.** −4   **9.** <   **10.** <   **11.** >   **12.** >   **13.** >   **14.** <   **15.** −15, −8, 2, 6
16. −5, 0, 7, 9   **17.** −9, 0, 3, 12

Objective 2.1B

1. –9   **2.** –3   **3.** 15   **4.** –7   **5.** 4   **6.** 5   **7.** –34   **8.** 28   **9.** –66   **10.** The opposite of negative eight
11. One plus negative six   **12.** The opposite of negative x **13.** Negative fifteen minus negative twelve
14. Negative four plus negative three   **15.** a minus b **16.** –3   **17.** –7   **18.** 5   **19.** 13
20. 4   **21.** 15

Objective 2.1C

1. 16   **2.** –19   **3.** –20   **4.** –65   **5.** 10   **6.** –3   **7.** 0.6   **8.** $2\frac{6}{7}$ **9.** –19   **10.** 28.1   **11.** $-\frac{5}{8}$ **12.** –9.7

13. <   **14.** <   **15.** >   **16.** <   **17.** =   **18.** >   **19.** –3, –1, 4, 6   **20.** –9, 3, 7, 10   **21.** –10, –6, 2, 5

Objective 2.1D

1. –16°C   **2.** 1°C   **3.** –2°C   **4.** –23°C   **5.** 0.74   **6.** –6.27
7. –8 (or 8 under par)   **8.** 3 (or 3 over par)   **9.** second quarter   **10.** third quarter

Objective 2.2A

1. 8	**2.** –2	**3.** –6	**4.** –27	**5.** –41	**6.** 88	**7.** 3
8. –9	**9.** –8	**10.** –14	**11.** –11	**12.** –13	**13.** 10	**14.** 17
15. 20	**16.** –12	**17.** –19	**18.** –10	**19.** –52	**20.** 0	**21.** 44
22. Yes	**23.** No	**24.** No				

Objective 2.2B

1. 4	**2.** –11	**3.** –21	**4.** 79	**5.** –16	**6.** –203	**7.** –12
8. 15	**9.** 19	**10.** –74	**11.** 14	**12.** 45	**13.** –5	**14.** –4
15. –20	**16.** 4	**17.** 24	**18.** 19	**19.** 50	**20.** –8	**21.** 2
22. Yes	**23.** Yes	**24.** No				

Objective 2.2C

1. 0°C	**2.** –11°C	**3.** 59 points	**4.** 131 points	**5.** –7°F	**6.** 28°F	**7.** 42°F
8. 27°F	**9.** 9261 meters	**10.** 5670 meters				

Objective 2.3A

1. –32	**2.** –56	**3.** 36	**4.** 99	**5.** 0	**6.** 493	**7.** –384
8. –640	**9.** 0	**10.** –78	**11.** 52	**12.** 120	**13.** –115	**14.** –72
15. –126	**16.** 21	**17.** –504	**18.** 324	**19.** No	**20.** No	**21.** Yes

Objective 2.3B

1. –3	**2.** –9	**3.** 7	**4.** –5	**5.** 12	**6.** undefined	**7.** –47
8. 0	**9.** 29	**10.** –12	**11.** –84	**12.** 70	**13.** –14	**14.** 4
15. –70	**16.** 15	**17.** –8	**18.** 7	**19.** Yes	**20.** No	**21.** Yes

Objective 2.3C

1. 2°C	**2.** –4°C	**3.** –12 points	**4.** 11 points	**5.** –3°	**6.** –1°	**7.** 14 shots
8. 20 shots	**9.** +6 points	**10.** –5 points				

Objective 2.4A

1. 12	**2.** 10	**3.** 14	**4.** 20	**5.** −7	**6.** −16	**7.** −16
8. 0	**9.** 0	**10.** −2	**11.** −7	**12.** −3	**13.** −6	**14.** −2
15. 12	**16.** 4	**17.** −13	**18.** 0	**19.** 0	**20.** 1	**21.** −20
21. 3						

Objective 2.4B

1. 37	**2.** −19	**3.** −16	**4.** −6	**5.** 4°C	**6.** $550	**7.** $15 million
8. $117 million						

Objective 2.5A

1. 10	**2.** −1	**3.** −38	**4.** −1	**5.** 5	**6.** 20	**7.** −20
8. −13	**9.** −16	**10.** −13	**11.** −16	**12.** 3	**13.** −12	**14.** −2
15. Undefined	**16.** 41					

CHAPTER 3

Objective 3.1A

1. 12	**2.** 21	**3.** 18	**4.** 40	**5.** 8	**6.** 36	**7.** 36	**8.** 30	**9.** 48	**10.** 75	**11.** 224	**12.** 36
13. 216		**14.** 252		**15.** 128		**16.** 63		**17.** 108		**18.** 231	**19.** 72
20. 200		**21.** 84	**22.** 154		**23.** 224		**24.** 80		**25.** 216	**26.** 36	**27.** 32

Objective 3.1B

1. 1	**2.** 3	**3.** 1	**4.** 2	**5.** 5	**6.** 5	**7.** 15	**8.** 4	**9.** 9	**10.** 7	**11.** 5	**12.** 10	**13.** 16
14. 13	**15.** 1	**16.** 2	**17.** 1	**18.** 3	**19.** 1	**20.** 5	**21.** 8	**22.** 4	**23.** 8	**24.** 9	**25.** 14	**26.** 17 **27.** 15

Objective 3.1C

1. GCF	**2.** GCF	**3.** LCM	**4.** LCM	**5.** 4 pairs	**6.** 20 copies
7. every 12 min	**8.** 30 minutes				

Objective 3.2A

1. $\frac{5}{8}$	**2.** $\frac{5}{7}$	**3.** $2\frac{4}{5}$	**4.** $2\frac{1}{2}$	**5.** $3\frac{1}{2}$	**6.** $3\frac{7}{8}$	**7.** $2\frac{3}{8}$	**8.** $2\frac{5}{8}$
9. $3\frac{2}{3}$	**10.** 6	**11.** $1\frac{8}{9}$	**12.** 3	**13.** $1\frac{5}{6}$	**14.** 1	**15.** $\frac{11}{5}$	**16.** $\frac{23}{3}$	**17.** $\frac{43}{9}$
18. $\frac{43}{4}$	**19.** $\frac{97}{10}$	**20.** $\frac{77}{6}$				

Objective 3.2B

1. 26	**2.** 15	**3.** 32	**4.** 5	**5.** 21	**6.** 18	**7.** 15
8. 36	**9.** 12	**10.** 72	**11.** 54	**12.** 120	**13.** $\frac{2}{3}$	**14.** $\frac{3}{5}$
15. $\frac{1}{8}$	**16.** $\frac{4}{7}$	**17.** $\frac{2}{3}$	**18.** $\frac{5}{8}$	**19.** $\frac{5}{8}$	**20.** $\frac{1}{2}$	**21.** $\frac{7}{18}$
22. $\frac{1}{4}$	**23.** $\frac{7}{12}$	**24.** 3				

Answers to Drill-and-Practice Pages

Objective 3.2C

1. < 2. > 3. > 4. < 5. < 6. > 7. > 8. > 9. < 10. < 11. < 12. > 13. > 14. <
15. < 16. < 17. > 18. < 19. < 20. > 21. < 22. > 23. < 24. < 25. > 26. < 27. >

Objective 3.2D

1. $\frac{7}{8}$ 2. $\frac{3}{8}$ 3. $\frac{1}{4}$ 4. $\frac{2}{3}$ 5. $\frac{1}{13}$ 6. $\frac{2}{3}$ 7. Yes
8. less

Objective 3.3A

1. $\frac{8}{13}$ 2. $\frac{6}{11}$ 3. $\frac{4}{27}$ 4. $\frac{49}{90}$ 5. $2\frac{7}{8}$ 6. $\frac{16}{35}$ 7. $\frac{13}{20}$
8. $\frac{15}{56}$ 9. 2 10. $2\frac{1}{2}$ 11. 25 12. $\frac{13}{30}$ 13. $1\frac{10}{21}$ 14. 0
15. $55\frac{1}{3}$ 16. $59\frac{1}{2}$ 17. $38\frac{1}{2}$ 18. 36 19. $1\frac{1}{27}$ 20. $\frac{15}{68}$ 21. $\frac{3}{22}$

Objective 3.3B

1. $\frac{1}{2}$ 2. $\frac{1}{2}$ 3. $\frac{5}{7}$ 4. 0 5. $1\frac{3}{8}$ 6. $\frac{8}{35}$ 7. $2\frac{5}{22}$
8. $\frac{2}{15}$ 9. $\frac{1}{2}$ 10. $\frac{4}{9}$ 11. Undefined 12. $\frac{2}{3}$ 13. $1\frac{1}{2}$ 14. $\frac{16}{55}$
15. $\frac{2}{5}$ 16. $1\frac{17}{21}$ 17. $1\frac{1}{4}$ 18. $1\frac{17}{40}$ 19. 2 20. $2\frac{15}{26}$ 21. $3\frac{1}{12}$
22. $\frac{3}{10}$ 23. $\frac{17}{33}$ 24. $3\frac{2}{3}$

Objective 3.3C

1. $174 2. 119 miles 3. 20 miles 4. 48 questions 5. $2\frac{3}{16}$ feet 6. $1408

7. 66,000 registered voters 8. $12 9. 258 pages 10. $5\frac{11}{16}$ miles 11. $624

12. 44 miles per hour

Objective 3.4A

1. $\frac{5}{6}$ 2. $1\frac{1}{10}$ 3. $\frac{10}{19}$ 4. $\frac{13}{15}$ 5. $1\frac{1}{2}$ 6. $4\frac{1}{4}$ 7. $1\frac{1}{12}$
8. $\frac{37}{45}$ 9. $1\frac{13}{24}$ 10. $1\frac{16}{21}$ 11. $1\frac{1}{42}$ 12. $1\frac{21}{40}$ 13. $1\frac{17}{36}$ 14. $1\frac{23}{24}$
15. $2\frac{4}{315}$ 16. $1\frac{9}{16}$ 17. $2\frac{13}{90}$ 18. $1\frac{4}{5}$ 19. $1\frac{13}{40}$ 20. $\frac{59}{165}$ 21. $2\frac{1}{3}$
22. $12\frac{1}{6}$ 23. $10\frac{7}{10}$ 24. $11\frac{6}{35}$ 25. $15\frac{7}{12}$ 26. $20\frac{13}{70}$ 27. $23\frac{7}{36}$

Objective 3.4B

1. $\dfrac{1}{6}$
2. $\dfrac{5}{12}$
3. $\dfrac{1}{3}$
4. $\dfrac{3}{7}$
5. $\dfrac{14}{27}$
6. $\dfrac{2}{13}$
7. $\dfrac{68}{135}$

8. $\dfrac{31}{80}$
9. $\dfrac{1}{78}$
10. $\dfrac{7}{20}$
11. $\dfrac{9}{35}$
12. $\dfrac{3}{10}$
13. $\dfrac{11}{72}$
14. $\dfrac{1}{78}$

15. $\dfrac{7}{40}$
16. $9\dfrac{15}{26}$
17. $4\dfrac{1}{16}$
18. $7\dfrac{3}{5}$
19. $18\dfrac{3}{7}$
20. $21\dfrac{5}{7}$
21. $3\dfrac{3}{5}$

22. $13\dfrac{29}{45}$
23. $34\dfrac{1}{48}$
24. $8\dfrac{21}{25}$
25. $9\dfrac{1}{6}$
26. $2\dfrac{17}{30}$
27. $126\dfrac{15}{56}$

Objective 3.4C

1. $3\dfrac{1}{8}$ inches
2. $1\dfrac{1}{16}$ inches
3. $5\dfrac{7}{24}$ feet
4. $9\dfrac{1}{4}$ pounds
5. $3\dfrac{7}{8}$ inches
6. $\$23\dfrac{1}{8}$

7. $19\dfrac{3}{8}$ long and $14\dfrac{5}{8}$ inches wide
8. $3\dfrac{1}{4}$ pounds
9. $\dfrac{7}{12}$ inch
10. $1\dfrac{7}{9}$ feet

Objective 3.5A

1. 40
2. -35
3. 25
4. $\dfrac{11}{21}$
5. $1\dfrac{2}{63}$
6. $-\dfrac{7}{8}$
7. $1\dfrac{2}{3}$

8. $-1\dfrac{3}{4}$
9. $1\dfrac{1}{2}$
10. $\dfrac{3}{25}$
11. $-\dfrac{7}{10}$
12. $-\dfrac{3}{7}$
13. $1\dfrac{5}{6}$
14. $\dfrac{1}{55}$

15. -12
16. $-1\dfrac{1}{4}$
17. $-\dfrac{1}{14}$
18. $\dfrac{4}{13}$
19. -5
20. $-2\dfrac{2}{3}$
21. $\dfrac{2}{19}$

Objective 3.5B

1. $\dfrac{8}{15}$
2. $1\dfrac{1}{9}$
3. -5
4. $-\dfrac{15}{28}$
5. 25 liters
6. 56,784 eligible voters
7. 544 miles
8. 481 miles

Objective 3.6A

1. $\dfrac{25}{121}$
2. $\dfrac{36}{49}$
3. $\dfrac{1}{64}$
4. $-\dfrac{8}{27}$
5. $11\dfrac{1}{9}$
6. $1\dfrac{11}{25}$
7. $\dfrac{9}{35}$

8. $\dfrac{1}{49}$
9. $-\dfrac{1}{18}$
10. $\dfrac{49}{120}$
11. $6\dfrac{1}{4}$
12. $-\dfrac{1}{15}$
13. $\dfrac{16}{625}$
14. $\dfrac{3}{40}$

15. $1\dfrac{1}{3}$

Objective 3.6B

1. $\dfrac{3}{8}$
2. $\dfrac{11}{25}$
3. 18
4. $\dfrac{2}{3}$
5. $\dfrac{8}{27}$
6. 1
7. $\dfrac{3}{4}$

8. 1
9. $1\dfrac{1}{8}$
10. $1\dfrac{2}{3}$
11. 6
12. $\dfrac{1}{8}$
13. $12\dfrac{1}{2}$
14. $6\dfrac{1}{4}$

Objective 3.6C

1. $\frac{43}{84}$ 2. $\frac{41}{45}$ 3. $\frac{7}{8}$ 4. $1\frac{89}{105}$ 5. $\frac{7}{8}$ 6. $\frac{11}{16}$ 7. $\frac{103}{252}$

8. $\frac{23}{90}$ 9. $\frac{4}{9}$ 10. $1\frac{19}{22}$ 11. $\frac{25}{36}$ 12. $\frac{2}{5}$ 13. 0 14. $1\frac{3}{98}$

15. $\frac{2}{3}$ 16. $1\frac{2}{23}$ 17. $2\frac{7}{8}$ 18. $\frac{1}{36}$ 19. 1 20. $\frac{5}{18}$ 21. $\frac{1}{3}$

22. $\frac{5}{21}$ 23. $\frac{9}{88}$ 24. $1\frac{19}{90}$ 25. $\frac{11}{36}$ 26. $4\frac{1}{3}$ 27. $2\frac{1}{8}$

CHAPTER 4

Objective 4.1A

1. 0.07 2. 0.353 3. 0.9 4. $\frac{501}{1000}$ 5. $\frac{7}{100}$ 6. $\frac{8411}{10,000}$

7. Thirty-nine hundredths 8. Eighty-one hundredths 9. Two and seven thousandths
10. Twenty-six and three hundred seventy-nine thousandths
11. Five hundred fourteen and three thousand one hundred eighteen ten-thousandths
12. One thousand seventy-eight and two hundred-thousandths 13. 0.834 14. 0.000052 15. 6.0101
16. 87.906 17. 25.07293 18. 91.0017

Objective 4.1B

1. < 2. < 3. > 4. > 5. > 6. < 7. < 8. < 9. > 10. < 11. > 12. > 13. <
14. > 15. > 16. < 17. 0.00037, 0.0037, 0.037, 0.37 18. 0.0086, 0.086, 0.851, 0.86 19. 0.049, 0.05, 0.49, 0.5
20. 0.0001, 0.012, 0.11, 0.21

Objective 4.1C

1. 0.1 2. 9.1 3. 26.3 4. 96.5 5. 65.34 6. 13.01 7. 517.68
8. 792.25 9. 2.092 10. 6.280 11. 79.463 12. 51.004 13. 0.0420 14. 0.0036
15. 4.3763 16. 16.1119 17. 0.24967 18. 0.00912 19. 7.88010 20. 11.73241 21. 1
22. 4 23. 71 24. 0.004590

Objective 4.1D

1. 19 mm 2. 1.6 mm 3. 2.27 g 4. Alphabet blocks 5. Emmitt Smith 6. Dan Marino
7. 41,000 yards 8. 3.3 ft 9. 1.15 miles 10. Billings, Montana

Objective 4.2A

1. 205.2844 2. 51.1918 3. 32.377 4. 5.421 5. 120.1664 6. 89.5197 7. 136.8126
8. 593.587 9. 61.5488 10. 9.3676 11. 2.3191 12. 23.126 13. 277.9922 14. 113.069
15. 62.446 16. 2.756 17. 53.8362 18. 106.8338 19. 93.3431 20. 0.062 21. 0.597
22. 3.091

Objective 4.2B

1. 36.3 miles 2. 2651.8 miles 3. $3.17 4. 25° 5. 0.7 pounds 6. 55.38 seconds
7. $1,062.72 8. $140,750.80

Answers to Drill-and-Practice Pages

Objective 4.3A

1. 0.35	**2.** 1.92	**3.** 50.4	**4.** 0.168	**5.** 22.95	**6.** 3.552	**7.** 0.0792
8. 0.04664	**9.** 0.2747	**10.** 10.58	**11.** 0.07062	**12.** 0.04712	**13.** 3.45285	**14.** 0.019154
15. 4.736	**16.** 0.456	**17.** 1.716	**18.** 82.1	**19.** 6823.5	**20.** 15.0788	**21.** 19.6935
22. 0.02142						

Objective 4.3B

1. 0.94	**2.** 3.2	**3.** 50	**4.** 86.5	**5.** 1.303	**6.** 0.31	**7.** 2.8
8. 470	**9.** 9.3	**10.** 7.7	**11.** 0.9	**12.** 29.4	**13.** 0.28	**14.** 19.51
15. 0.69	**16.** 0.080	**17.** 0.831	**18.** 0.008	**19.** 1	**20.** 16	**21.** 70
22. 0.723	**23.** 3.15	**24.** 0.626				

Objective 4.3C

1. $0.\overline{8}$ **2.** 0.214   **3.** $0.\overline{54}$ **4.** 4.375   **5.** 41.3   **6.** 1.087   **7.** $0.9\overline{4}$

8. $6.\overline{5}$ **9.** 10.571   **10.** $\dfrac{37}{50}$ **11.** $\dfrac{3}{8}$ **12.** $\dfrac{41}{200}$ **13.** $4\dfrac{69}{500}$ **14.** $6\dfrac{8}{125}$

15. $3\dfrac{7}{20}$ **16.** $\dfrac{1}{9}$ **17.** $4\dfrac{81}{100}$ **18.** $\dfrac{11}{200}$ **19.** $<$ **20.** $<$ **21.** $>$

22. $>$ **23.** $<$ **24.** $<$

Objective 4.3D

1. 37.5 inches	**2.** 113 miles	**3.** $6996	**4.** $2744	**5.** $107.06	**6.** 836 gallons	**7.** 23 hairdryers
8. 6.2 minutes						

Objective 4.4A

1. 2.52	**2.** 5.09	**3.** 6.3	**4.** −28.27	**5.** −21.73	**6.** 10.03	**7.** −0.4
8. −2.1	**9.** −2.52	**10.** 20.1	**11.** 20.77	**12.** 0.88	**13.** −3.185	**14.** −0.125
15. −13	**16.** −0.4					

Objective 4.4B

1. 116.8 ft/s	**2.** $7.9 million	**3.** $16.50	**4.** $374.50	**5.** 16 in.	**6.** 13.5 ft

Objective 4.5A

1. 7   **2.** −4   **3.** 12   **4.** −11   **5.** 7   **6.** 9   **7.** 13

8. 5   **9.** −36   **10.** 19   **11.** 23   **12.** $\dfrac{5}{8}$ **13.** $\dfrac{2}{7}$ **14.** $\dfrac{2}{63}$

15. $\dfrac{9}{20}$ **16.** 6   **17.** −20   **18.** 72   **19.** 9

Objective 4.5B

1. 4.1231	**2.** 29.6985	**3.** −48.4665	**4.** −84.2496	**5.** 20.3961	**6.** −11.1355	**7.** 28.7228
8. 42.6028	**9.** −51.4393	**10.** 12, 13	**11.** 8, 9	**12.** 10, 11	**13.** $2\sqrt{6}$	**14.** $4\sqrt{5}$
15. $5\sqrt{5}$	**16.** $4\sqrt{10}$	**17.** $2\sqrt{70}$	**18.** $10\sqrt{3}$	**19.** $6\sqrt{7}$	**20.** $5\sqrt{2}$	**21.** $3\sqrt{14}$

Objective 4.5C

1. 33 ft/s	**2.** 42 ft/s	**3.** 27 ft/s	**4.** 2.75 sec	**5.** 5 sec	**6.** 4 sec

Objective 4.6A

1.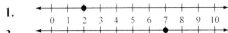
2.
3.
4.
5.
6.
7.
8.
9.
10.
11.
12.
15.
16.

Objective 4.6B

1. 3 **2.** $-7.6, -3, 0$ **3.** $-6, 1.9$ **4.** $0, 6$ **5.** $-5, -1.1$ **6.** 10

7.
8.
9.
10.
11.
12.
13.
14.

Objective 4.6C

1. $s \geq 75,000$; no **2.** $c < 220$; yes **3.** $h \leq 9$; no **4.** $w > 1850$; yes **5.** $b \leq 2250$; yes

6. $p > 80$; no

CHAPTER 5

Objective 5.1A

1. $10x$ **2.** $24x$ **3.** $-6a$ **4.** $12y$ **5.** $56y$ **6.** $-12x^2$ **7.** $-63x^2$

8. x^2 **9.** x^2 **10.** a **11.** x **12.** x **13.** a **14.** a

15. x **16.** x **17.** y **18.** n **19.** $3x$ **20.** $4x$ **21.** $-4x$

22. $-3x$ **23.** $-28a^2$ **24.** $-15x^2$ **25.** 0 **26.** 3 **27.** m

Objective 5.1B

1. $-x-5$ **2.** $6x-3$ **3.** $8x-12$ **4.** $-3a-6$ **5.** $-4a-48$

6. $-4y+10$ **7.** $-12y+16$ **8.** $-6x^2-36$ **9.** $-20x^2-8$ **10.** $-10y^2+15$

11. $-21y^2+70$ **12.** $-6x^2-6y^2$ **13.** $-5x^2+5y^2$ **14.** $-3x^2+6y^2$ **15.** $-10a^2+16b^2$

16. $6x^2+6x-42$ **17.** $3x^2-6x-15$ **18.** $-2y^2-6y+10$ **19.** $-3a^2-3a-12$ **20.** $14x^2-28x-42$

21. $6x^2-8x+2$ **22.** $2x^2+4xy-8y^2$ **23.** $6x^2-9xy-12y^2$ **24.** $-2a^2-4a+3$ **25.** $9b^2+5b-8$

Objective 5.2A

1. $11x$ **2.** $21x$ **3.** $5a$ **4.** $7a$ **5.** $2ab$ **6.** $5xy$ **7.** $-2xy$

8. $-4xy$ **9.** $5ab$ **10.** 0 **11.** $-\dfrac{7}{12}x$ **12.** $-\dfrac{2}{5}x$ **13.** $-\dfrac{1}{3}x^2$ **14.** $\dfrac{1}{8}x$

15. $-\dfrac{17}{12}x^2$ **16.** $8a$ **17.** $-13x^2$ **18.** $-y^2$ **19.** $x+2y$ **20.** $7y-5x$ **21.** $7x$

22. $6a-12b$ **23.** $7y-11x$ **24.** $-2x^2-x$ **25.** $-5b+14a$

Answers to Drill-and-Practice Pages

Objective 5.2B

1. $-x-21$ **2.** $-a-1$ **3.** $13-10x$ **4.** $17-6x$ **5.** $-2-5y$ **6.** $7n-6$ **7.** $4x-10$

8. $-2x+26$ **9.** $x-17$ **10.** $4y+2$ **11.** $12y-33$ **12.** $-x+5y$ **13.** $2a+5b$ **14.** $-6x+18$

15. $20x+90$ **16.** $3x-18$ **17.** $4x-48$ **18.** $-4x+16$ **19.** $10x+5$ **20.** $-5x+30$ **21.** $22x-48$

22. $-4x+49$ **23.** $a+2b$ **24.** $8b$ **25.** $13x-24y$ **26.** $22x-y$

Objective 5.3A

1. $-x^2+2x$ **2.** $5y^2+3y$ **3.** y^2+y-5 **4.** $2x^2+7x-12$ **5.** $4x^2+10x-8$

6. y^2-5y-6 **7.** $3x^2+8x+11$ **8.** $4x^2+3x+3$ **9.** $-2y^2-9$ **10.** y^3+2y^2-2y-3

11. $3a^3-10a^2+7a$ **12.** $2y^3+2y^2-6y-2$ **13.** $5x^2+6x$ **14.** y^2+8y

15. $7x^2+xy-2y^2$ **16.** $7x^2+4xy-3y^2$ **17.** $-2x^2-9x-7$ **18.** $4a^2-3a+15$ **19.** $-5x^2+5x+1$

20. $5a^2-6a+11$ **21.** $-5x^2+2x-1$ **22.** $4x^3+9x^2-x-2$ **23.** $5y^3+2y^2-8y+9$ **24.** $6x^3+x^2-x-8$

Objective 5.3B

1. $4x$ **2.** $-6y$ **3.** $3y^2-3y-6$ **4.** $-9x-5$ **5.** x^2-2x-9

6. $-x^2+12x+6$ **7.** $3x^3+2x^2+3x+1$ **8.** $-2x^3-4x^2-2x$ **9.** $3x^3+2x^2+3x+1$ **10.** $-x^2-x-1$

11. $4y^3+2y^2+4y+2$ **12.** x^3-4x-1 **13.** $-3xy$ **14.** $3x^2-5xy$ **15.** y^2-2xy

16. $9y^2+y+7$ **17.** $2x^2-3x+4$ **18.** $-3x^2+7x+1$ **19.** $-x^3+2x^2+7$

20. $3x^3-2x^2-8x-5$ **21.** $2x^3+x^2+3x+6$ **22.** $3x^3-8x^2+10x-3$ **23.** $-3x^2+4x+1$

24. $-2a^3+a^2-a+6$

Objective 5.3C

1. $(8x-1)$ meters **2.** $8x$ inches **3.** $(5x^2-x-3)$ km **4.** $(8y^2+9y-3)$ mi

5. $(-0.1n^2+235n-850)$ dollars **6.** $(-0.3n^2+305n-2000)$ dollars **7.** $(-0.7n^2+125n-1500)$ dollars

8. $(-0.5n^2+600n-600)$ dollars

Objective 5.4A

1. $3x^2$ **2.** $-2y^2$ **3.** $6x^2$ **4.** $16y^5$ **5.** $8a^5$ **6.** $-8a^{10}$ **7.** x^3y^5

8. $-12x^5y$ **9.** $-9a^6b^2$ **10.** a^3b^5 **11.** $-8x^3y^5$ **12.** $5a^3b^5$ **13.** x^3y^5z **14.** $a^3b^3c^3$

15. x^4y^4z **16.** $-a^4b^7$ **17.** $-x^4y^4$ **18.** $28y^7z^4$ **19.** $-6x^4y^2$ **20.** $a^4b^4c^6$ **21.** $x^4y^8z^3$

22. $24a^6b^6$ **23.** $-24x^5y^7$ **24.** $-6a^7b^6$ **25.** $8x^5y^8$ **26.** $-6a^{10}b^4$ **27.** $-30a^4b^3c^3$

Objective 5.4B

1. x^4 **2.** x^6 **3.** y^8 **4.** x^{10} **5.** y^{12} **6.** x^6 **7.** $8x^3$

8. $9y^2$ **9.** $-8x^3$ **10.** $9y^4$ **11.** x^2y^4 **12.** x^8y^{12} **13.** $-8a^3b^6$ **14.** $x^{10}y^{10}$

15. x^5y^{15} **16.** $x^{16}y^{12}$ **17.** $4x^4y^6$ **18.** $-a^6b^{12}$ **19.** a^6b^6 **20.** $-54y^7$ **21.** a^4b^3

22. $-12x^5y^4$ **23.** $-18x^4y^3$ **24.** $4a^3b^8$ **25.** $-54a^5b^6$ **26.** $-24a^6b^5$ **27.** $-243a^5b^8$

Objective 5.5A

1. $x^2 + x$ 2. $2y - y^2$ 3. $-x^2 - 2x$ 4. $2a^2 - 2a$ 5. $3b^2 + 15b$ 6. $-2x^3 + 2x^2$

7. $-4y^3 - 24y^2$ 8. $-2x^4 + 3x^2$ 9. $10x^3 - 4x^2$ 10. $6y^2 - 3y^3$ 11. $2x^2 - 3x$ 12. $6x^2 - 3x$

13. $-x^3y + x^2y^3$ 14. $-2x^2y^2 + xy^3$ 15. $x^3 - 2x^2 + x$ 16. $x^3 + 3x^2 - 2x$ 17. $-a^3 + 6a^2 + a$

18. $-2b^3 - 3b^2 + 6b$ 19. $2x^4 - 3x^3 - 2x^2$ 20. $-2y^5 - 3y^4 + 4y^3$ 21. $-4y^4 - 10y^3 + 16y^2$

22. $12x^4 - 6x^3 + 21x^2$ 23. $20x^4 - 4x^3 - 36x^2$ 24. $-5y^4 + 15y^3 + 30y^2$ 25. $a^3b - 3a^2b^2 - 4ab^3$

26. $x^3y - 2x^2y^2 + 2xy^3$ 27. $a^3b + 5a^2b^2 - 7ab^3$

Objective 5.5B

1. $y^2 + 5y + 6$ 2. $a^2 + 3a - 10$ 3. $b^2 - b - 20$ 4. $y^2 - 5y - 14$ 5. $x^2 + 5x - 36$

6. $y^2 - 8y + 12$ 7. $a^2 - 10a + 16$ 8. $x^2 + 8x - 33$ 9. $2x^2 + 13x + 6$ 10. $3y^2 + 5y + 2$

11. $2x^2 + 3x - 9$ 12. $5x^2 + 13x - 6$ 13. $6x^2 - 13x + 5$ 14. $2y^2 + 7y - 9$ 15. $4y^2 + y - 14$

16. $30a^2 + 17a + 2$ 17. $12a^2 - 56a + 65$ 18. $10a^2 - 53a + 63$ 19. $15b^2 - 43b - 44$

20. $12a^2 + 31a - 30$ 21. $x^2 + 3xy + 2y^2$ 22. $2a^2 + 5ab + 2b^2$ 23. $2x^2 - 5xy + 3y^2$

24. $2a^2 - 3ab - 9b^2$ 25. $8a^2 + 18ab - 5b^2$ 26. $9x^2 - 9xy - 10y^2$ 27. $30x^2 + 17xy + 2y^2$

Objective 5.6A

1. $\dfrac{1}{36}$ 2. 1 3. $\dfrac{1}{125}$ 4. $\dfrac{1}{a^5}$ 5. $\dfrac{x}{y^2}$ 6. 1 7. $\dfrac{1}{a^{14}}$

8. $\dfrac{y^2}{x^2}$ 9. $\dfrac{1}{x^4}$ 10. 1 11. $3x$ 12. $-3x^3$ 13. $\dfrac{b^6}{a^2}$ 14. x^4y^3

15. $3y$ 16. $\dfrac{a^2}{25}$ 17. x^6 18. -1 19. $\dfrac{-3a^2b^4}{4}$ 20. $\dfrac{b^2}{-6a}$ 21. $\dfrac{-2y^2}{11z^3}$

22. $\dfrac{x^4y^3}{4}$ 23. $-b^2c$ 24. $\dfrac{x}{y^3}$

Objective 5.6B

1. 6.75×10^{-10} 2. 1.15×10^9 3. 5.635×10^{-7} 4. 5.6×10^6 5. 9.761×10^{10} 6. 4.06×10^{13} 7. 5.96×10^{-7}

8. 1.906×10^{-5} 9. 0.000034 10. $180,500,000,000,000$ 11. $60,950,000$ 12. 0.000000034

13. 0.00000169 14. 0.0000000209 15. $456,000$ 16. $960,000,000,000$

Objective 5.7A

1. $6 + x$ 2. $y - 10$ 3. $12 - x$ 4. $t + 6$ 5. $a - 1$ 6. $y - 5$ 7. $\dfrac{z}{8}$

8. $11m$ 9. $x^2 + 10$ 10. $15 - x^3$ 11. $5(n + 8)$ 12. $\dfrac{2}{3}x + 6$ 13. $x + 3x$ 14. $\dfrac{-6}{y}$

15. $-2t$ 16. $2(x + 8)$ 17. $5(y + 6)$ 18. $\dfrac{10}{y + 5}$ 19. $\dfrac{1}{4}y^2 + 12$ 20. $-1m + 16$ 21. $x + \dfrac{x}{2}$

22. $\dfrac{3}{5}(10w)$ 23. $y - 2y$ 24. $7(x + 3)$

Answers to Drill-and-Practice Pages

Objective 5.7B

1. $4n - n$; $3n$
2. $5n + n$; $6n$
3. $(n+12) - 3$; $n+9$
4. $n + (11-n)$; 11
5. $n + (n+7)$; $2n+7$
6. $n - (n+5)$; -5
7. $\frac{2}{5}n + \frac{3}{10}n$; $\frac{7}{10}n$
8. $\frac{3}{8}n - \frac{1}{4}n$; $\frac{1}{8}n$
9. $(n+9) - 3$; $n+6$
10. $(n+3) + 4$; $n+7$
11. $4(5n+6)$; $20n+24$
12. $11n - 7n$; $4n$
13. $(n+10) + 12$; $n+22$
14. $15 - (n+5)$; $10 - n$
15. $3(n+n+1)$; $6n+3$
16. $(n+n+1) + 9$; $2n+10$
17. $\frac{1}{2}(n+n+2)$; $n+1$
18. $(n-9) + (n+20)$; $2n+11$
19. $5n - n$; $4n$
20. $6(n^2+4)$; $6n^2+24$

Objective 5.7C

1. gallons of paint in first container: g; gallons of paint in the second container: $10 - g$
2. length of first piece: L; length of second piece: $35 - L$
3. speed of corporate jet: s; speed of commercial jet: $3s$
4. amount at 3.5%: A; amount at 4%: $4200 - A$
5. height of triangle: h; base of triangle: $h + 4$
6. width: w; length: $4w$
7. length of shorter piece: L; length of longer piece: $8 - L$
8. width of rectangle: w; length of rectangle: $2w - 3$
9. number of apple trees: N; number of cherry trees: $\frac{1}{5}N$
10. number of full-time faculty: F; number of adjunct faculty: $F + 206$

CHAPTER 6

Objective 6.1A

1. 5
2. 3
3. 15
4. 14
5. 9
6. −5
7. −8
8. 2
9. 3
10. 4
11. 2
12. 0
13. −5
14. 2
15. −6
16. 2
17. 6
18. −8
19. 10
20. 15
21. −6
22. 16
23. −1
24. $\frac{1}{4}$
25. $\frac{3}{5}$
26. $-\frac{5}{6}$
27. $-\frac{7}{6}$

Objective 6.1B

1. 3
2. 6
3. −5
4. −5
5. −3
6. 12
7. 7
8. 6
9. −7
10. 3
11. −6
12. −4
13. 12
14. 20
15. −12
16. −15
17. 14
18. −20
19. −16
20. 8
21. −6
22. −30
23. 12
24. 16
25. $\frac{3}{2}$
26. 2
27. 5

Objective 6.2A

1. 6
2. 7
3. 4
4. 7
5. −3
6. −5
7. −7
8. 2
9. 2
10. 1
11. 5
12. 6
13. 2
14. 2
15. 7
16. −4
17. −7
18. 4
19. 9
20. 2
21. 2
22. 3
23. 1
24. 2
25. 6
26. 4
27. −9

Objective 6.2B

1. 60 cm
2. 24 cm
3. $96
4. 42 months
5. 12,320 cm^3
6. 15 cm
7. $1200
8. $11,500
9. 180 cm^2
10. 20 cm

Answers to Drill-and-Practice Pages

Objective 6.3A

1. 7	**2.** 4	**3.** 8	**4.** 5	**5.** 4	**6.** −2	**7.** −3
8. −2	**9.** −3	**10.** 2	**11.** −3	**12.** −2	**13.** −2	**14.** −4
15. −6	**16.** 0	**17.** 0	**18.** −1	**19.** −2	**20.** 2	**21.** −1
22. −3	**23.** 6	**24.** 10	**25.** −2	**26.** −3	**27.** 6	

Objective 6.3B

1. 2	**2.** 3	**3.** −6	**4.** 7	**5.** −1	**6.** 1	**7.** −2
8. 3	**9.** 2	**10.** −2	**11.** −7	**12.** 6	**13.** 8	**14.** −3
15. 0	**16.** −1	**17.** −12	**18.** 2	**19.** −3	**20.** 4	**21.** −3
22. 2	**23.** 2	**24.** 9	**25.** −4	**26.** −2	**27.** 0	

Objective 6.3C

1. 6 ft from the 36-lb force **2.** 6 ft from the 50-lb force **3.** 48 lb **4.** 8 lb **5.** $35 **6.** $15
7. $200 **8.** $70 **9.** 51.25°C **10.** 60.8°C

Objective 6.4A

1. 6	**2.** 9	**3.** 16	**4.** 18	**5.** 2	**6.** 3	**7.** 5
8. −12	**9.** −3	**10.** 4	**11.** 4	**12.** −5	**13.** 4	**14.** 7, 14
15. 6	**16.** 14					

Objective 6.4B

1. $14	**2.** $80	**3.** $3200	**4.** $27,000	**5.** 960	**6.** 80°, 40°	**7.** 2 h
8. 40 mph	**9.** 60 cm, 30 cm, and 90 cm	**10.** 1450 rpm				

Objective 6.5A

1.   **2.**   **3.**   **4.**

5. 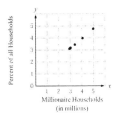 **6.**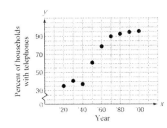

7. *A* is (2, 0), *B* is (−3, 1), *C* is (1, 3), *D* is (−2, -3) **8.** *A* is (−4, 1), *B* is (−3, −2), *C* is (3, 2), *D* is (1, 4)

Objective 6.5B

1.

2.

143

Answers to Drill-and-Practice Pages

Objective 6.6A

1. Yes   **2.** No   **3.** No   **4.** Yes   **5.** Yes   **6.** No   **7.** Yes
8. No   **9.** No   **10.** Yes   **11.** (2, 3)   **12.** (−3, −7)   **13.** (4, 0)   **14.** (10, 1)
15. (−3, 7)   **16.** (5, −3)   **17.** (−4, 3)   **18.** (6, −3)

Objective 6.6B

1. 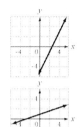   **2.**    **3.** 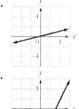   **4.**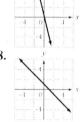

5. **6.** **7.** **8.**

CHAPTER 7

Objective 7.1A

1. 1.96 m   **2.** 24.3 cm   **3.** 4218 m   **4.** 963 cm   **5.** 345 mg   **6.** 720 g   **7.** 0.254 g
8. 2.754 kg   **9.** 2.347 kl   **10.** 2.7 L   **11.** 3310 L   **12.** 3.2 ml   **13.** 270 kg   **14.** 8 VCRs
15. 120 kg   **16.** 560.52 L   **17.** $0.385   **18.** 75 L

Objective 7.2A

1. $\frac{3}{7}$, 3:7, 3 to 7   **2.** $\frac{3}{8}$, 3:8, 3 to 8   **3.** $\frac{1}{7}$, 1:7, 1 to 7   **4.** $\frac{3}{2}$, 3:2, 3 to 2

5. $\frac{9}{10}$, 9:10, 9 to 10   **6.** $\frac{5}{2}$, 5:2, 5 to 2   **7.** $\frac{18}{7}$ **8.** $\frac{4}{9}$ **9.** $\frac{91 \text{ miles}}{3 \text{ gallons}}$ **10.** $\frac{\$62}{3 \text{ toasters}}$

11. $\frac{1 \text{ tablet}}{6 \text{ hours}}$ **12.** $\frac{3 \text{ clams}}{1 \text{ person}}$ **13.** $\frac{8 \text{ place settings}}{1 \text{ table}}$ **14.** $\frac{217 \text{ seats}}{2 \text{ lecture halls}}$ **15.** 8.4 gallons/minute
16. 68 heartbeats/minute   **17.** 64 words/minute   **18.** $1250/partner   **19.** 124 calls   **20.** 57 words

Objective 7.3A

1. 112 in.   **2.** 5 ft   **3.** 14 ft   **4.** 10 yd   **5.** 198 in.   **6.** $2\frac{2}{3}$ yd   **7.** 15,840 ft

8. 81 in.   **9.** 2 lb   **10.** 10,000 lb   **11.** 4 tons   **12.** 1200 lb   **13.** 224 oz   **14.** $3\frac{1}{5}$ tons

15. $5\frac{5}{8}$ lb   **16.** 156 oz   **17.** 3 c   **18.** $3\frac{1}{2}$ pt   **19.** 5 c   **20.** 5 qt   **21.** 3 gal

22. 24 qt   **23.** 13 pt   **24.** 16 c

Objective 7.3B

1. $1\frac{1}{8}$ ft   **2.** 126 ft   **3.** 22.5 ft   **4.** 750 lb   **5.** 48 lb   **6.** 27 lb   **7.** 22 qt

8. 7 gal   **9.** $3\frac{3}{4}$ c   **10.** 21 pt

144

Objective 7.3C

1. 200.2 m **2.** 85 kg **3.** 22.74 L **4.** 707 ml **5.** 96.6 km/h **6.** \$3.28/kg **7.** \$3.18/L
8. 3542 km **9.** 2.12 gal **10.** 8200 ft **11.** 17.64 oz **12.** 16.89 gal **13.** 1.57 in. **14.** 1.92 pt
15. 47.20 mph **16.** \$1.95/lb

Objective 7.4A

1. True **2.** True **3.** True **4.** Not true **5.** True **6.** Not true **7.** True
8. True **9.** Not true **10.** 6 **11.** 5 **12.** 12 **13.** 24 **14.** 108
15. 10 **16.** 33.6 **17.** 6 **18.** 24.75 **19.** 17.5 **20.** 7.5 **21.** 4.5
22. 2.14 **23.** 1.2 **24.** 30

Objective 7.4B

1. \$90.40 **2.** 10.5 teaspoons **3.** \$31.05 **4.** 34 feet **5.** \$1036 **6.** \$113.70
7. 2259 transistors **8.** \$9.73 **9.** 10,380 people **10.** 434 gallons

Objective 7.5A

1. 6 **2.** 7 **3.** 35 **4.** 90 **5.** 42 **6.** 8 **7.** \$322
8. 21 lbs **9.** 1152 words **10.** 333.2 ft **11.** 2048 ft **12.** 5.2 h

Objective 7.5B

1. 60 **2.** 6 **3.** 112 **4.** 7 **5.** 27 **6.** 30 **7.** 27 ft
8. 60 mph **9.** 1445 computers **10.** 40 revolutions **11.** 56.25 lumens **12.** 115.2 lbs

CHAPTER 8

Objective 8.1A

1. $\frac{39}{100}$, 0.39 **2.** $\frac{16}{25}$, 0.64 **3.** $1\frac{1}{4}$, 1.25 **4.** $\frac{13}{50}$, 0.26 **5.** $\frac{17}{20}$, 0.85 **6.** $\frac{1}{5}$, 0.20 **7.** $4\frac{1}{2}$, 4.50

8. $\frac{19}{100}$, 0.19 **9.** $\frac{11}{20}$, 0.55 **10.** $\frac{71}{900}$ **11.** $\frac{23}{300}$ **12.** $\frac{129}{500}$ **13.** $\frac{129}{200}$ **14.** $\frac{13}{30}$

15. $\frac{399}{400}$ **16.** 0.675 **17.** 0.3407 **18.** 0.579 **19.** 0.40 **20.** 0.1389 **21.** 0.0201

Objective 8.1B

1. 32% **2.** 96% **3.** 4% **4.** 197% **5.** 214% **6.** 0.9% **7.** 68%
8. 12% **9.** 10.7% **10.** 41.7% **11.** 188.9% **12.** 57.1% **13.** 15.6% **14.** 60%

15. 183.3% **16.** $45\frac{5}{11}$% **17.** $22\frac{2}{9}$% **18.** $114\frac{2}{7}$% **19.** $37\frac{1}{2}$% **20.** $6\frac{2}{3}$% **21.** $58\frac{1}{3}$%

Objective 8.2A

1. 3.6 **2.** 15 **3.** 18.2 **4.** 47.7 **5.** 27 **6.** 19.25 **7.** 88.075
8. 12.688 **9.** 112 **10.** 0.05 **11.** 45 **12.** 0.2 **13.** 147 **14.** 13.2
15. 64.8 **16.** 5.25 **17.** 9.48 **18.** 11 **19.** 0.3 **20.** 36 **21.** 192.5
22. 38.75

Objective 8.2B

1. 289   **2.** 60   **3.** 30%   **4.** 35%   **5.** 60   **6.** 200   **7.** 50%
8. 25%   **9.** 510   **10.** 800   **11.** 37%   **12.** 450%   **13.** 768   **14.** 819
15. 60   **16.** 150   **17.** 40%   **18.** 52%   **19.** 5%   **20.** 12.5%   **21.** 350%
22. 400%   **23.** 16,500   **24.** 2500

Objective 8.2C

1. 486 plants   **2.** 16.2 gallons   **3.** $27,560   **4.** $188,000   **5.** 12%   **6.** 12%   **7.** 20%
8. 3%   **9.** 25%   **10.** 80%   **11.** 75%   **12.** 12.5%   **13.** 97.5%   **14.** 7 errors

Objective 8.3A

1. 25%   **2.** 8.3%   **3.** 33.3%   **4.** 10%   **5.** 150 fans   **6.** 230 billboards
7a. $13.50   **7b.** $238.50   **8a.** $2560   **8b.** $34,560   **9a.** 50 spaces   **9b.** 5%
10a. 300 water meters   **10b.** 4300 water meters   **11.** 12.5%   **12.** 15%

Objective 8.3B

1. 15%   **2.** 30%   **3.** 12%   **4.** 15%   **5.** 81 employees   **6.** $375
7a. 6 minutes   **7b.** 37.5%   **8a.** 49 orders   **8b.** 35%   **9a.** $17,100   **9b.** $267,900
10a. $63   **10b.** $357   **11.** 24%   **12.** 5%

Objective 8.4A

1. $7.80   **2.** $0.34   **3.** $10.88   **4.** 60%   **5a.** $13.60   **5b.** $47.60   **6a.** $19.80
6b. $64.80   **7a.** $0.60   **7b.** $1.69   **8a.** $9.10   **8b.** $35.10   **9.** $64.86   **10.** $44.64

Objective 8.4B

1. 36%   **2.** 15%   **3.** $30   **4.** $75   **5.** 20%   **6.** $4.80   **7a.** $0.60
7b. $1.80   **8a.** $9.90   **8b.** $45.10   **9a.** $4.20   **9b.** 12%   **10a.** $120   **10b.** 20%
11. $138   **12.** $39

Objective 8.5A

1. $1800   **2.** $8960   **3.** $99,200   **4.** $14,790   **5.** $6225   **6.** $5.40   **7a.** $714
7b. $2814   **8a.** $23,400   **8b.** $88,400   **9a.** $56,250   **9b.** $131,250   **10a.** $3608   **10b.** $11,808

CHAPTER 9

Objective 9.1A

1. $180°$ **2.** $64°$ **3.** $15°$ **4.** $68°$ **5.** $52°$ **6.** $41°$ **7.** $24°$
8. $42°$ **9.** $138°$ **10.** $78°$ **11.** $9°$ **12.** $9°$ **13.** $149°$ **14.** $113°$
15. $63°$ **16.** $116°$ **17.** $84°$ **18.** $268°$

Objective 9.1B

1. $12°$ **2.** $14°$ **3.** $47°$ **4.** 5   **5.** $a = 123°, b = 57°$ **6.** $a = 39°, b = 141°$
7. $20°$ **8.** $25°$

Objective 9.1C

1. 50° **2.** 37° **3.** 55° **4.** 69° **5.** 14° **6.** 29° **7.** 30°
8. 57°

Objective 9.2A

1. 54 cm **2.** 53.5 cm **3.** 37.68 cm **4.** 88 in. **5.** 27 ft **6.** 33 ft **7.** 62 m
8. 15 ft 11 in. **9.** 139 cm **10.** 145 ft **11.** 21,120 ft **12.** 251.2 ft **13.** 376.8 ft **14.** 62 ft
15. 20 ft **16.** 50.24 in.

Objective 9.2B

1. 7 ft^2 **2.** 35.2 cm^2 **3.** 49 ft^2 **4.** 841 cm^2 **5.** 1404 cm^2 **6.** 325 in.^{2} **7.** 379.94 in.^{2}
8. 60.84 cm^2 **9.** 405 cm^2 **10.** 975 cm^2 **11.** 254.34 ft^2 **12.** 452.16 in.^{2} **13.** 65,111.04 ft^2 **14.** 54,000 m^2

Objective 9.3A

1. 10 in. **2.** 7.211 cm **3.** 9 ft **4.** 8.246 in. **5.** 12.806 ft **6.** 14.142 yd **7.** 125.377 m
8. 9.849 ft **9.** 26 mi **10.** 24 km

Objective 9.3B

1. $\frac{1}{3}$ **2.** $\frac{3}{4}$ **3.** 4 cm **4.** 2.9 cm **5.** 3.3 in. **6.** 11.3 cm **7.** 33 ft **8.** 24 ft
9. 15 cm **10.** 3375 cm^2

Objective 9.3C

1. Yes, SSS Rule **2.** Yes, SAS Rule **3.** No **4.** Yes, ASA Rule **5.** Yes, SAS Rule
6. Yes, ASA Rule **7.** Yes, SSS Rule **8.** No

Objective 9.4A

1. 6600 cm^3 **2.** 53,240 cm^3 **3.** 144.70 ft^3 **4.** 10,648 cm^3 **5.** 254.04 ft^3 **6.** 904.32 mm^3 **7.** 7234.56 cm^3
8. 11.46 m^3 **9.** 11.46 m^3 **10.** 18 m^3 **11.** 282.6 m^3 **12.** 27,468.72 ft^3 **13.** 381.5 m^3 **14.** 3052.08 ft^3
15. 416 m^3 **16.** 2304 ft^3

Objective 9.4B

1. 168.54 cm^3 **2.** 615.75 in^2 **3.** 200π in^2 **4.** 1400 m^2 **5.** 6.5 ft **6.** 28π in^2 **7.** 5 cans
8. 589.05 cm^2

CHAPTER 10

Objective 10.1A

1. 31 **2.** 1–4: 5; 5–8: 5; 9–12: 8; 13–16: 5; 17–21: 5; 22–24: 4; 25–28: 4; 29–32: 4
3. 9–12 **4.** 12 **5.** 13 **6.** 13 **7.** 12.5% **8.** 87.5%

Objective 10.1B

1. 20 cars **2.** 5 cars **3.** 50 cars **4.** 10% **5.** 25 employees **6.** $\frac{1}{4}$

7. 60 employees **8.** 50 employees

Answers to Drill-and-Practice Pages

Objective 10.1C

1. 25 families 2. $\frac{5}{21}$ 3. 30 families 4. $\frac{4}{7}$ 5. 30 homes 6. $\frac{1}{5}$ 7. 50 homes 8. 70 homes

Objective 10.2A

1. $16.32 2. 88 3. $271 4. 232 pizzas 5. $36.75 6. 103 mi 7. $8.38
8. 22 9. no mode 10. 93

Objective 10.2B

1. 24 years 2. 46 years 3. 26 years 4. 36 years 5. 29 years 6. 22 years 7. 10 years
8. 75 9. 150 10. 75 11. 75% 12. 0%

Objective 10.2C

1. 4.472 2. 1.414 3. 1.512 4. The standard deviations are the same.

Objective 10.3A

1. $\frac{1}{12}$ 2. $\frac{1}{9}$ 3. 1 4. $\frac{3}{8}$ 5. $\frac{4}{11}$ 6. $\frac{2}{11}$
7. Drawing a diamond 8. $\frac{2}{13}$ 9. $\frac{1}{3}$ 10. $\frac{1}{3}$

Objective 10.3B

1. 1 to 3 2. $\frac{1}{35}$ 3. $\frac{5}{31}$ 4. $\frac{12}{1}$ 5. $\frac{3}{1}$ 6. $\frac{4}{7}$ 7. $\frac{17}{1}$
8. $\frac{1}{21}$